THE LITERATURE OF DEATH AND DYING

This is a volume in the Arno Press collection

THE LITERATURE OF DEATH AND DYING

See last pages of this volume for a complete list of titles

DEATH

ITS CAUSES AND PHENOMENA

HEREWARD CARRINGTON

ARNO PRESS

A New York Times Company

New York / 1977

Reprint Edition 1977 by Arno Press Inc.

THE LITERATURE OF DEATH AND DYING
ISBN for complete set: 0-405-09550-3
See last pages of this volume for titles.

Manufactured in the United States of America

Library of Congress Cataloging in Publication Data

Carrington, Hereward, 1880-1959.
Death, its causes and phenomena.

(The Literature of death and dying)
Reprint of the 1921 ed. published by Dodd, Mead, New York.
1. Death (Biology) 2. Immortality. I. Title. II. Series.
QP87.C3 1977 616.07 76-19563
ISBN 0-405-09559-7

DEATH

ITS CAUSES AND PHENOMENA

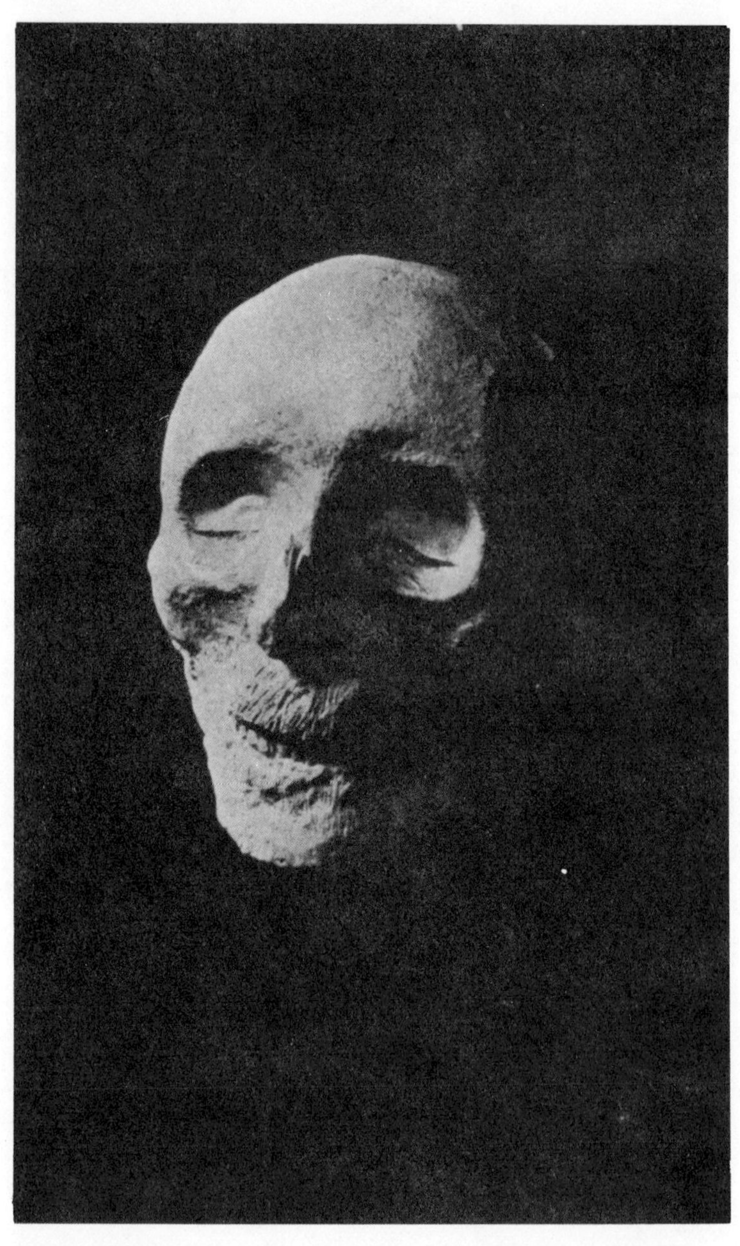

Head of Richelieu in the Church of the Sorbonne, Paris, from a photograph taken in 1897, two hundred and seventy-three years after his death.

Frontispiece.

DEATH
ITS CAUSES AND PHENOMENA

WITH SPECIAL REFERENCE
TO IMMORTALITY

BY

HEREWARD CARRINGTON, Ph.D.

Author of "Vitality, Fasting and Nutrition," "The Coming Science," "The Physical Phenomena of Spiritualism," "Modern Psychical Phenomena," "Your Psychic Powers," etc.

WITH FRONTISPIECE

NEW YORK
DODD, MEAD AND COMPANY
1921

VAIL-BALLOU COMPANY
BINGHAMTON AND NEW YORK

PREFACE

DEATH is that last "Great Adventure" which we must all experience.

A tragedy for those left behind—those who are still living—death is usually considered a tragedy for the one who has died; yet this it cannot be!

Death is a solemn and a terrible thing, but that is no reason why we should be afraid of it. Rather, we should regard it as the great deliverer from a life which might otherwise become unbearable, and when this is not the case, it proves, merely, that such a death has been *premature.* As such, it is not *death itself* which is terrible; it is the premature termination of an otherwise useful or beautiful life.

The subject discussed at length in this volume—Death—is generally looked upon as something to be "tabooed" by polite society; something unpleasant, which may some day come upon us, but which we desire to think about as little as possible in the interval. There is no logical ground for this position, however, and, scientifically speaking, death may be made as fascinating a study as any other. Divested of the superstition and glamour which usually surround it, death assumes the appearance of a most interesting scientific problem, both from its physiological and from its psychological aspects.

There is also another side to this question which must by no means be overlooked. I refer to the possibility of postponing death, on the one hand, and of rendering it more painless, on the other. Both of these results can only be effected by a thorough understanding of the process involved; and this, in turn, can only be obtained by a close scientific study of the problem—one that includes all its aspects, and treats of them impartially. In summing up this evidence, in condensing what has been said—the speculations that have been offered during the past two hundred years (*see* Bibliography)—I believe that I have collated a considerable amount of interesting material; while the particular theories as to the *nature* of death which have been advanced, will not, I hope, be without interest, and perhaps utility.

The present edition of this work is considerably abridged from the English edition, which appeared some ten years ago. The material relating to Psychical Research, in particular, I have condensed into a single Chapter. On the other hand, I have endeavoured to bring Part I—dealing with the Physiological aspects of the subject—up-to-date, so as to include the latest researches in this direction. The English edition appeared over the joint signature of Mr. J. R. Meader and the present writer, but since then, Mr. Meader has himself solved the Problem with which this book deals. It has therefore been necessary for me to revise it myself, to the best of my ability. Several Chapters of the original edition—embodying Mr. Meader's own views—have, therefore, been omitted; but I have endeavoured to preserve the bulk of the material *in statu quo,* merely adding such newer evidence as later researches seemed to warrant.

H. C.

CONTENTS

PART I—PHYSIOLOGICAL

CONTENTS

PART II—PSYCHOLOGICAL

PART I

PHYSIOLOGICAL

DEATH

CHAPTER I

THE SCIENTIFIC ASPECT OF LIFE AND DEATH

DEATH is universally recognized as the inevitable fate of every living thing—the goal towards which animate life is constantly tending—and yet, strange as it may appear, human ingenuity has not yet succeeded in formulating a definition that will adequately cover this last experience of man. We know that all things that live must grow old and die, but our theories concerning the causes that produce this phenomenon are still almost entirely of a speculative character. To say that death "is a cessation of life" is to avoid the question. Even Spencer's definition, in which he pronounced life to be "the continual adjustment of internal to external relations," and death, a want of correspondence between those relations, leaves much to be desired. It presents the *facts* of life and death as we behold them, but it fails absolutely to trace these apparent effects to the *causes,* of which they are the natural manifestation.[1]

As far as positive science is concerned, the only immortality that can be demonstrated is that of race. The individual dies, from natural causes or by accident, as the case may be, but, as each living thing is the direct result

[1] "Is it not obvious that this definition merely gives or states the ***effects*** of life—its phenomena—and does nothing to state what its real essence is at all? . . . Life is ***that which adjusts,*** not the adjustments themselves."—***Vitality, Fasting and Nutrition,*** pp. 334–5.

of reproduction from another form, the death of the individual has practically no effect upon the continuance of existence of the race. With this so-called potential immortality, therefore, science is satisfied. Beyond this it finds no room for certitude—no opportunity for its experiments.

To make this position clear to the mind of those who have not been accustomed to the materialistic view of the phenomena of life and death, it may be necessary to explain that science recognizes no new organism in the product of reproduction any more than it distinguishes a new creation in the changes that are so constantly occurring in the form of living matter. Even a slight acquaintance with the first principles of science is sufficient to explain what this means, for we know that the atoms that constitute the human body are so lacking in stability that they are ever being discarded and replaced by other substances derived through the process of assimilation. In other words, the one property that best distinguishes living matter from dead matter is what might be termed the faculty of self-creation, or the ability to transform the dead substances assimilated into the same live substance of which this matter is composed. Thus, as long as life continues, this process goes on with unceasing regularity. Dead matter is cast aside, just as one would discard a worn-out garment, and new matter is created to take its place. When this faculty ceases to perform its functions, death follows speedily.

Both Huxley and Cuvier have used the river whirlpool as an exact illustration of the nature of this phenomenon of life, and most physiologists agree that this whirl of water, as seen, for example, at Niagara, is an extremely close reproduction of the natural process of assimilation

and disintegration—the alternating attraction and re
sion of the ever-changing particles representing
actual conditions of physical life. That a material sub-
stratum is left unchanged, there can be no doubt; but
even this theory does not modify the conclusions that
science has drawn from this reproduction of the whirl
of life. Though it may be true that the animal body contains permanent elements of definite composition, they
alone are insufficient to assure the continuance of physical existence.

It seems to be the popular impression that this physical body begins its work of development at birth; that
it continues to progress until the individual has attained that rather indefinite period generally termed
"maturity," and that, when this point has been reached,
definite deterioration commences. This idea has been
seriously questioned, and many practical experiments in
biology indicate that the body begins to lose its re-
creative powers, or the capacity to change dead matter
into living matter, very shortly after the period of birth,
and that, from this time, the decrease in force continues
steadily. As one writer has said:—

"In want of a more exact knowledge of the structure of the
living molecule and the changes in structure that come on in
old age, the physiologist expresses his idea of the general nature of these changes by similes and metaphors more or less apt.
We may compare living matter to a clock, the mainspring of
which is so constructed that, in consequence of slowly developing molecular changes, it suffers a gradual loss of elasticity.
In such a mechanism there will come a time when 'winding
the clock' will no longer make it run, since energy can no longer
be stored in the spring. We may imagine this loss of elasticity
to develop gradually, giving stages that may be roughly com-

pared to the periods of life. To carry out the simile, it is the food we eat and the oxygen we breathe that take the place of the winding force. In consequence of a slowly developing molecular change in the organism, this energy is less efficiently utilized as the individual grows older. The clock runs more feebly and needs relatively more frequent winding, until at last the elasticity is gone, the power of assimilation is insufficient, and we have what we call natural death."[1]

Brown, in his article on "Old Age,"[2] has expressed this view more briefly. "The causes of death," he says, "are not to be found in the summation of many external injuries, but are already established within the organism itself, and death is simply the natural end of development." If this theory be true, it is very contradictory to the definition formulated by Spencer in his *Principles of Biology*. The latter would logically lead the student to conclude that "external relations" play the most important part in determining the length of life, and that, if perfect correspondence between the internal and external relations could be secured, existence would continue interminably. As has been shown, however, this idea is entirely contrary to the beliefs of modern physiologists. In their opinion, man would still die, even though there were no injurious changes of environment, as the natural weakening of the assimilative powers would alone be sufficient to make death inevitable.

Of course the simile of the clock is too simple an illustration to be applied comprehensively to so complex an organism as the human body. In this combination of living matter there is no single mainspring to wear out—

1 W. H. Howell in *Reference Handbook of Medical Sciences.*
2 *British Medical Journal,* Oct. 3, 1891.

no one cause of death against which man may protect himself—and it is due to these conditions that death does not come to every portion of the body at precisely the same moment. While it is necessarily true that death is actually the cessation of the normal functions upon which life depends, the causes which result in the suspension of the bodily mechanism may arise in any one of the several important or vital centres. According to the arrangement devised by Bichat, death may be divided into three classes:—(1) that which begins at the heart; (2) that which begins at the lungs; and (3) that which begins at the head. But the collapse of the vital force in a single one of these centres is sufficient to bring death with more or less rapidity to every other portion of the organism.

But, while most physiologists hold that it is the ultimate fate of all living things to die, it must not be imagined that this is the only theory to which scientists subscribe, for there are some biologists who are inclined to accept Weismann's speculative conclusions, as presented in his *Essays upon Heredity*.

In these papers, this eminent biologist expresses the opinion that all living matter once possessed potential immortality, and that death is a condition that came into the world because of the continued existence of the individual had assumed the proportions of a serious danger to the general well-being of the species. In other words, death is a condition that did not necessarily exist in the beginning of things, but was eventually adopted for the reason that just such a safety-valve was necessary to permit of the perpetuation of the race.

As an illustration in proof of this theory, Weismann

draws our attention to the amoeba, one of the unicellular organisms or protozoa, which biologists recognize as the lowest forms of animal life. While a complete cell in itself, performing all the functions of assimilation and reproduction, it knows no process of dissolution that can be compared to the phenomenon that we designate as death. On the contrary, its very act of reproducing its species is, in itself, a striking example of the possibility of "physical immortality," for it is the fate of this creature to continue to increase in size until, finally, the limit of growth is reached. At this point the original cell divides into two parts, and, where one organism existed, there are now two individuals, both of which are capable of performing the functions of life, and of dividing in turn into two cells—a process of reproduction that, so far as science has been able to ascertain, goes on indefinitely.

Of course, the objection may be raised—as it has been—that the original individual cell dies in the act of reproducing its offspring, and that the two cells that result from this physical separation of the larger body are actually different individualities. To this Weismann replied that there is no death in this change "because there is no corpse." In this fission we have the illustration of the continuance of life, not its dissolution, i. e., not its actual "death."

It is upon this hypothesis that Weismann bases his theory that living matter originally possessed the element of potential immortality, and he explains the appearance of death among the metazoa by reference to the law of natural selection.

If this theory be correct, the possibility of never-ending existence possessed by the unicellular creature was undoubtedly passed on to the more complex organism which, in the process of evolution, was eventually produced from this lowlier manifestation of animate life. In the course of time, however, certain new but important conditions arose. In the first place, death became a necessity to the perpetuation of the species; and, in the second place, the division of functions among the many cells of the metazoa made the immortality of each particular cell unnecessary for reproductive purposes.

The very name that has been applied to this law of evolution, "natural selection," gives an indication of the pitiless qualities that mark its operations. As its name implies, its tendency is always towards the promotion of the good of the race, without regard to the particular interests of the individual. Thus, when it became apparent that natural death was needed to remove those individuals who were not only no longer necessary to the welfare of the species, but were actually an adverse element or obstacle in the path of natural progress, the presence of those cells that were no longer required in the process of fecundity gave nature the opportunity to effect this adjustment in the laws governing the struggle for existence.

As students of biology are well aware, bodily structures that are of no further use to nature soon retrograde, or disappear almost completely. As an example, we have the cave-dwelling animals and fishes, which, despite the fact that they show every indication of having once had eyes, are now sightless. That is to say, when the time came that they had no further use for eyes, nature permitted the sense of sight to degenerate, and at last, even the physical organs themselves deteriorated, until only a rudimentary record was left of the member that had once actually existed.

In this illustration, Weismann finds an explanation of the process by which the element of immortality was lost by the many-celled organisms. Being not only of no further utility, but of positive danger to the species, its perpetuation would have retarded the realization of the purpose of evolution. Through the operation of the law of natural selection, therefore, death came as a beneficent solution to this great problem of the moment, the limitation of the production to those individuals who would be of service in helping to carry out the scheme of the perpetuation of the species.

It must be stated in this connection, however, that Weismann's theory is seriously questioned at the present day. Thus Haeckel, in his *Wonders of Life*, pp. 99–100, points out that:—

"The immortality of the unicellulars, on which Weismann has laid so much stress, can only be sustained for a small part of the protists even in his own sense—namely, for those which simply propagate by cleavage, the chromacea and bacteria among the monera, the diatomes and paulotomes among the protophyta, and a part of the infusoria and rhizopods among the protozoa. Strictly speaking, the individual life is destroyed when a cell splits into daughter cells. One might reply to Weismann, that in this case the dividing unicellular organism lives on as a whole in its offspring, and that we have no corpse, no dead remains of the living matter left behind. But that is not true of the majority of the protozoa. In the highly-developed ciliata the chief nucleus is lost, and there must be from time to time a conjunction of two cells and a mutual fertilization of their secondary nuclei before there can be any further multiplication by simple cleavage. However, in most of the sporozoa and rhizopoda, which generally propagate by spore formation, only one portion of the unicellular organism is used for this; the other portion dies, and forms a 'corpse.' . . . "

The fact is that each metazoon consists of many successive generations of cells—it really is a cell cycle—and can only be homologized with a cycle of protozoan generations, not with any single protozoan, which is but a single cell. Hence it follows that the death of an individual protozoan is not homologous with the death of an individual multicellular organism. Weismann committed the fundamental error of assuming the complete homology of the two forms of death, and thus reached the false conclusion that protozoa are all certainly potentially immortal.

E. Maupas contended that there is a distinct loss of vitality in protozoa in the course of successive generations, and that conjugation must occur at some stage to effect rejuvenescence. G. N. Calkins (*Studies in the Life-History of Protozoa*) takes the same view—that the development of the protozoa is *cyclical;* and this is supported by M. Hartmann, who contends that natural death does occur among the protozoa.

H. S. Jennings in his book *Life and Death, Heredity and Evolution in Unicellular Organisms,* has summarized the conclusions of modern science regarding the potential immortality of such organisms, up to the year 1920, and he concludes as follows:—

"As individuals the *infusoria* do not die, save by accident. Those that we now see under our microscopes have been living ever since the beginnings of life; they came from division of previously existing individuals. . . . The result of the work so far done has been to confirm the view that the *infusoria* may live indefinitely without mating. I believe that we may look upon this as one of the secure results of science. There are many of the unicellular creatures, particularly the bacteria, in which nothing like mating is known. . . . But in just the same sense it is true for ourselves that every one that is alive now has been alive since the beginning of life. This truth applies at least to our *bodies* that are alive now; every cell of all our bodies is a piece of one or more cells that existed earlier, and thus our entire body can be traced in an unbroken chain as far back into time as life goes. The difference is that in man and other higher organisms there have been left all along the way great masses of cells that did not continue to live. . . . From our own personal view it seems unfortunate that the mass of cells which is next to wear out and be left behind in the chain of life is that with which our own selves seem to be bound up; but certain samples of our selves may continue to live indefinitely, like the *infusorian.*"

Thus it will be seen that, while the useless part of life, so to say—the body—is cast off or discarded at death, the central core of life seems to be potentially immortal. It has existed, so far as we can see, from the beginnings of time in the past and will continue forever into the future. This conclusion arrived at by materialistic biological science is surely one of the profoundest significance and importance, and, taken in conjunction with the doctrine of the conservation of energy, seems to indicate very strongly the importance of life and its possible subsequent persistence—its "potential immortality." How far this doctrine may logically be applied to the possible persistence of life *apart* from the physical body remains to be seen.

Mr. Newman Smyth in his interesting book *The Place of Death in Evolution* has succinctly summarized the facts regarding the entrance of death into the cosmic scheme and its significance and value with regard to life. He says:—

> "The first fact which has been observed is that natural death does not appear immediately at the beginning of the history of life on the earth. . . . By isolating individual *infusoria*, and thus preventing them from renewing their power of reproduction by meeting other distantly related forms of their own species, Maupas discovered among the descendants of the isolated individual, increasing signs of enfeebled life, senescence, and the loss of power to multiply; and finally the succession of their generations came to a complete pause, and a dead cell was left at the end of it. *At this point natural death, so far as now known, first appeared.*"

It is to be noted however that, experimentally, this point is seldom attained. Sedgwick and Wilson indeed in their *General Biology* (p. 170) say:—"In nature, this limit is probably seldom, if ever, reached." It is interesting to note also that the rudiments of sex coincide largely with the beginnings of death upon our globe.

Death is now known to occupy a useful, indeed necessary place in the process of evolution. Without death evolution would have been impossible, for as life becomes increasingly more organized and complex, death prevails. It is necessitated by the increasing complexity of life and the differentiation of the original protoplasm into separate parts or organs having separate functions. It has frequently been pointed out that death is the price we pay for this increased complexity of structure. Death assists in the general scheme of creation; it has a great utility. Unless death had supervened, the more finely organized and fairer forms of life would not have appeared. The course of life would have been arrested. Man himself would not have appeared. The earth before us has died that we might live. We are the living children of a world that has died for us!

Death, therefore, was not introduced at the outset for the sake of annihilating life, but that it might help and hasten the progress of life, until it should reach its present point of comparative independence in our spiritual being. Only the simplest living creatures, which multiply by division are potentially immortal; but when multiplication results from some form of fertilization, with the consequent introduction of sex, this potential immortality of the whole mass is lost, and the death of the individual is necessarily introduced, in order to preserve and improve the species through evolution,—even at the expense of the individual organism. Viewed in this larger light, therefore, we must recognize the utility of death as a means for furthering life, and not for the purpose of destroying it. "I came not to destroy, but to fulfill."

When we come to speak of "death," moreover, we must be very sure that we understand our terms accurately, as much confusion has always arisen because of inaccurate definition in all the sciences no less than in philosophy and metaphysics. We must be very sure as to just what we mean by "death" before we can undertake to argue about it; and there are some very loose conceptions afloat which it would be well to check at the outset of the investigation. Let us see what these are.

When we cut off a chicken's head, we say that the chicken is "dead"; its conscious life is extinguished, and if it continues to move, or even to run about the yard, as it does sometimes, we do not assume for that reason that any "life" still remains in the chicken, but rather that "reflex action" causes these phenomena. On the other hand, if we pluck a rose, it keeps its freshness for several days, and, until that rose has withered and lost its freshness and beauty entirely, we do not say that the rose is "dead." In the one case, we assume that death has taken place instantaneously; in the other, that death does not take place for several days. Why is this?

The difficulty arists from this fact. There are in reality *two kinds of death*, which are confused in the public mind until only one death is recognized—a compound of these two. And yet, to keep the problem perfectly clear, it is very essential that these two kinds of death should be kept strictly apart, and in no wise confused. Only in that way can the problem be understood. Let us take the two instances we have given, and with them, as examples, see if we cannot make this problem somewhat clearer, and distinguish the two, so that there may be no more confusion upon this important point.

When the chicken's head was cut off, its conscious life came to a termination at that moment. It is probable that the subsequent movements *were* purely reflex, and not in any way the result of any conscious action and volition. The conscious life of the chicken ended at that moment therefore. *But the body, the cells, and tissues of the chicken did not die at that time.* The body of the chicken—the tissues—lived on for several hours, and not until the last remnant of vitality had departed could we say that the bird was "dead." That is to say, the tissues of the body continued to live on for several hours after the conscious life of the bird had ceased. This tissue or cell-life, the life of the body, is technically known as "somatic life," as distinct from conscious or mental life. Now, in the case of the rose, we do not as a rule say that it is "dead" until somatic death has taken place. It is probable that the "conscious" life of the rose *did* come to a termination at its plucking; at that moment its "conscious" life, so far as it can be said to have one, came to an end, while its *somatic* life did not. Since the rose does not show its mental life in the same way that a chicken does, however, it is very difficult to *prove* this fact, and doubtless many would contend that no such conscious life existed at all. It is a question almost incapable of proof, but it has always seemed probable that by analogy there must be some sort of conscious life that is terminated at the moment of the picking of the flower. At all events, these examples will help to clear up this problem, and enable us to distinguish the *two kinds of death*—the conscious and the somatic—which must be kept carefully in mind throughout the following discusssion.[1]

[1] A tissue is said to "die" when it loses permanently its power of responding to its appropriate stimuli. The brain and nervous system die, in man and warm-blooded animals, at the moment of somatic death; gland tissue dies very soon after. Smooth muscle retains its irritability forty-five minutes, skeletal muscle some hours, after death.

"The greater the degree of interdependence existing between the actions of its several parts, the more is the well-being of the entire organism interfered with by damage occurring to any one of these special parts. Through the intervention, for the most part, of the nervous system and vascular system, this individuality of the entire organism is carried to the most marked extent in the highest vertebrata, so that the life of one of these creatures—regarded as a whole, or sum-total of phenomena—differs almost as widely as it is possible from that of some of the lowest animals on the one hand, and from the plants on the other. *Their mode of death also is quite different.* And as with life, so with death; we are perhaps too apt to form our notions concerning each from what we see taking place in man himself and in the higher living things."—Bastian, *The Beginnings of Life*, vol. i, p. 108.

While science has, however, been unable to arrive at a positive conclusion regarding the origin or nature of death, it is by no means so difficult to determine the probable bounds or limitations to the duration of life. Omitting those instances which depend upon tradition for their verification, or that cannot be authenticated because of our inability to fix the unit of time used in making the calculations, or for any other reason, we occasionally find cases which show that the scriptural limitation of "threescore years and ten" falls far short of representing the greatest possible length of physical existence in man. Even today the death of a centenarian is not an unknown occurrence. At the same time, this question of human longevity is a much disputed one, and many facts have to be taken into consideration when estimating the evidential value of such cases, and particularly the historic cases. Leaving out of account, for the time being, the Biblical records, there are certain historical cases that have been quoted time and time again in proof of the possible limit of man's life; but these historic examples are, strangely enough, very rarely investigated. This is to be regretted, for such cases, almost without exception, when closely inquired into, are found to rest upon totally inadequate evidence. Mr. William J. Thoms investigated a number of such cases very minutely, going into the histories of the cases with extreme care, and published the results of his investigations in a book entitled, *Human Longevity; Its Facts and Its Fictions*, etc. The author shows us how careless statements are frequently the cause of mistakes that go for a hundred years or more before they are corrected, if indeed they ever are. Mr. Thoms also points out several sources of error, any of which might have vitiated the results in many instances. Mistaken identity may have taken place—two people of the same name having lived in a certain parish, etc. Again, a married couple may have a son who dies. They have a second son a number of years later, and they give this son the same name as the first child. These two get confused in memory and in record, and it is generally the second, or even the third and youngest son that lives to a "good old age"; and he, being confused with the first or second child of like name, becomes celebrated for being many years older than he really is.

A number of such sources of error are shown, and backed up by several cases in which these errors had doubtless taken place. The inaccuracy of baptismal certificates, tombstones, etc., is also illustrated. Mr. Thoms examined in detail the famous cases of Henry Jenkins, Thomas Parr, and the Countess of Desmond. Original trials, documents, army and navy registers, parish registers, etc., were examined in every instance.[1] The cases of Parr, Jenkins, and that of the Countess of Desmond, when examined, were found to be resting on quite inadequate proof; indeed, there was no proof at all, that could properly be called evidential! The author gives a number of carefully-investigated cases, the results of which are, briefly, as follows:—Mary Billings, reputed 112 years old, proved to be 91; Jonathan Reeves, 104, proved to be 80; Mary Downton, 106, proved to be 100; Joshua Millar, 111, proved to be 90; Maudit Baden, 106,

[1] Among other interesting documents in this connection, the reader may consult *Evidences of the Great Age of Henry Jenkins, with Notes, respecting Longevity and Long-Lived Persons.* Bell, Richmond, 1859. The case of old Thomas Parr (who was examined *post-mortem* by Harvey) is to be found in a work entitled, *The Olde, Olde, Very Olde Man; or, The Age and Long Life of Thomas Parr.*

proved to be considerably less,—how much less is not certain; Thomas Geeran, 106, *ditto;* John Pratt, 106, *ditto;* George Fletcher, 108, proved to be 92; George Smith, 105, proved to be 95; Edward Couch, 110, proved to be 95; William Webb, 105, proved to be 95; John Dawe, 108 or 116, proved to be 87; George Brewer, 106, proved to be 98; Robert Howlinson, 103, proved less; Robert Bowman, 118 or 119, proved much less; Frederick Lahrbush, 106, proved less; Richard Purser, 112, proved less; William Bennett, 105, proved to be 95; Mary Hicks, 104, proved to be 97; and several others. The author gives four cases, however, in which the ages 102, 100, 103, and 101 had undoubtedly been reached, and a chapter of cases in which ages of more than one hundred might possibly be presumed, although the evidence was not strong enough to prove the fact. But after the evidence adduced in the former portion of the book, it is certain that all such statements, especially if not backed up by documentary evidence, are to be mistrusted. Dr. De Lacy Evans gives some seventy cases of persons who had apparently reached an age of more than one hundred (*How to Prolong Life: An Inquiry into the Cause of "Old Age" and "Natural Death,"* etc., pp. 100–121, London, 1885); but none of his cases are well certified, and the names of several of the discredited cases figure prominently. The same may be said of the collection of forty-seven cases given by Dr. Hosmer Bostwick, in his *Inquiry into the Cause of Natural Death; or, Death from Old Age* (New York, 1851). There is no doubt, however, that certain cases of old age *do* sometimes come up. So far as I know, Captain Diamond's great age of 112 years has never been disproved. Metchnikoff gives us the portrait of an old woman of 105 years of age in his *Prolongation of Life* (p. 6); and it is stated, upon the authority of Albert Kruger, Superintendent of the Home of the Daughters of Jacob in New York City, that Mrs. Esther Davis, an inmate of this institution, was in 1908, 115 years of age.[1]

It is all the more astonishing that there should be so few trustworthy examples of old age, when we take into account the fact that it is all but universally conceded that from 100 to 120 years should be the normal limit of life of the individual man and woman. The fact that so few actually *do* reach this age, proves conclusively how perverted are the food and other habits of the people.

Although we know, therefore, both from experience and from authentic historical facts, that men and women do occasionally pass the centenary mark, it must be admitted that such cases are rather exceptional, for, so far as modern mortality statistics are concerned, the average length of human life is nowhere much in excess of forty-five years.

Strictly speaking, therefore, practically the only positive fact that science can teach us concerning death is that it is the inevitable fate of all living things. The law that stipulates that all those who are born must die is now as certain in its operation as the law of gravitation. At this point, however, materialistic science stops,

1 In his *Philosophy of Long Life* Jean Finot has given a number of cases in which men have lived much longer than a hundred years, and some of them an incredible time; but his cases do not seem to rest on any very secure basis—many of the old cases being quoted which Mr. Thoms has conclusively shown to be incorrect. At the same time, I admit that some of his cases seem well established, while others will be found in T. E. Young's little book *On Centenarians* (London, 1899). Voronoff, in his book on *Life,* has again quoted several of the older (discredited) cases.

leaving the probable fate of the individuality, or thinking-part of man, an unsolved problem. As to this "soul-part" of being, in fact, science has even questioned its very existence. To the ordinary scientist, death is a door which closes upon consciousness as the breath leaves the body. If there be any existence behind that door, his experiments have thrown no light upon it, and the man who is unwilling to accept these negative conclusions as the last word on this subject must search elsewhere for the evidence in support of the hope that is within him.

CHAPTER II

THE MOMENT OF DEATH

1. The Hour of Death

That more deaths occur at some particular hour of the day or night than at any other time has been more than once maintained by statisticians, who have always produced figures to support their claims. A recent essay in this line, the investigations of Dr. H. D. Marsh, of New York, indicate that the wave of diurnal efficiency, or the range of mental and physical activity, varies with the habits of the individual as regards work and sleep, and that with inhabitants of civilized communities the hour of greatest efficiency is likely to be 5 P. M. Says *The British Medical Journal* (London, January 18, 1910), in an article on this subject:—

"This conclusion was the outcome of a special investigation conducted by Dr. Marsh, and curiously enough an examination of the records of death in New York City, likewise made by him, showed that during the period under examination 5 P. M. was also the hour at which the majority of 23,439 deaths from disease occurred. It is certainly notable that the period of the twenty-four hours at which the average man is most alive should be the same as that at which his death is most likely to occur, and the apparent inconsistency has led to turning over our own columns in search of previous observations on the question of what may be called the hour of death. The general result is to indicate that before any final statement can be made as to the hour of the twenty-four at which *caeteris paribus* death is most likely to occur in any given individual, much more extended and thorough investigations of the point will have to be carried out than have yet been undertaken. At present the evidence is somewhat conflicting. Thus it is found that Finlayson, writing in the Glasgow *Medical Journal* and using some statistics compiled by the City Chamberlain, found that of 13,000 deaths recorded in 1865, the greatest number occurred between the hours of 5 and 6 A. M., while Schneider, writing in *Virchow's Archiv* on deaths in Berlin, concluded that the most fatal hour was between 4 A. M. and 7 A. M. The number of deaths upon which he based his conclusions was 57,000; while Berens, arguing from the limited number of 1000 deaths in Philadelphia, and writing in the Philadelphia *Medical Times*, concluded in favour of the hour between 6 A. M. and 7 A. M. In 1896 Dr. C. F. Beadles published the result of an examination of the statistics of Colney Hatch Asylum. These showed a difference between the two sexes as regards the hour of greatest mortality. Thus, among 1000 women the most fatal hour was between 6 and 7 in the evening, while among 3424 men it was between 5 and 6 in the morning."

Apparently there is a pretty wide choice here for those who prefer to die at the popular hour. The majority, however, would appear to lean toward the earlier hours of the day, as against the conclusion reached by Dr. Marsh. None of them, however, the writer in *The British Medical Journal* reminds us, give countenance to the popular belief that an invalid is most likely to succumb at about 2 A. M., when, according to the Duke of Wellington, the heroic attitude is most difficult to assume. We read in conclusion:—

"On the surface of things, it seems unlikely that any particular hour should be more fatal than another, and in any case it is clear that those who have investigated the matter have not always been dealing with truly comparable units. Precision in recording the exact hour of death is not easy to obtain, and, besides this, data such as the nature of the illness, its duration, and the age and sex of the patient, have also to be considered. The observers, as a rule, seem to be alive to this point."

2. Pain at the Moment of Death

Contrary to general opinion, there is seldom any pain at the moment of death. A great deal of evidence could be adduced in support of this statement, but I shall content myself with citing a certain number of authorities and a limited amount of evidence only. Dr. Thomas D. Spencer, writing in the *Popular Science Monthly,* some years ago, said:—

"At birth the babe undergoes an ordeal that, were he conscious, would be more trying than a most painful death; yet he feels it not. Born in an unconscious state, the brain incapable of receiving conscious impressions, his entrance into this hitherto unknown world is accomplished during a state of oblivion, known as 'Nature's anaesthesia.' From the earliest period of history, death has been considered as necessarily accompanied by pain; so general is this belief, that the terms 'death agony,' 'last struggle,' 'pangs of death,' etc., have been in almost universal use in every age and under all conditions of society.

"Nothing could be more erroneous; the truth is, pain and death seldom go together—we mean the last moments of life. Of course, death may be preceded by weeks or even months of extreme suffering, as occurs during certain incurable diseases.

"The blood sent to the brain is not only diminished in quantity, but is laden with carbonic- acid gas, which, acting on the nervous centres, produces a gradual benumbing of the cerebral ganglia, thereby destroying both consciousness and sensation. The patient gradually sinks into a deep stupor, the lips become purple, the face cold and livid, cold perspiration (death-damp) collects on the forehead, a film creeps over the cornea, and, with or without convulsions, the dying man sinks into his last sleep. As the power of receiving conscious impressions is gone, his death struggle must be automatic. . . . Even in those cases where the senses are retained to the last, the mind is usually calm and collected, and the body free from pain."

Professor Tyndall stated that death by lightning must be quite painless, and, from an experience of his own, in which he was shocked into insensibility, on one occasion, he should be entitled to speak upon this point with exceptional authority (*Fragments of Science*). Dr. Edward Clark, in his book on *Visions,* asserted that "death is no more painful than birth." Dr. James M. Peebles stated that in all cases of death from shock, there could be no pain—consciousness being obliterated too suddenly. Henry Ward Beecher asserted that "there is no pain at the last moment." An article in the *Medical National Review,* some years ago, pointed out that death, in cases where a rifle ball passes through the brain, etc., must be painless. Many other cases and statements to like effect could be adduced, if it were necessary.[1]

Of course there is pain in a certain number of cases; of that there can be no doubt. In a few cases, notably in those who "fight for life," self-consciousness, with pain, *is* present, but such cases are very rare. In most cases, "nature's anaesthetic" is doubtless operative, and there is no pain.

[1] See MacKenna, *The Adventure of Death;* Mercer, *Why Do We Die?* etc., for late testimony upon this point.

Regarding this question of pain at the moment of death, Dr. Osler has said:—

"I have careful records of about five hundred death-beds, studied particularly with reference to the modes of death and the sensations of the dying. The latter alone concern us here. Ninety suffered bodily pain or distress of one sort or another, eleven showed mental apprehension, to positive terror, one expressed spiritual exaltation, one bitter remorse. The great majority gave no sign one way or the other; like their birth, their death was a sleep and a forgetting." [1]

Says M. Finot [2]: —

"The pains which accompany death are chiefly imaginary. Even putting on one side accidental death caused by the breakage of nerves, apoplectic strokes, and diseases of the heart, in which pain is absent, the cases in which we suffer at the approach of death are very rare."

It is a curious fact that pain is generally lost when nature "gives up the fight." In cases of cancer, *e. g.* pain is experienced so long as there is life and activity, but this pain almost invariably passes away a few hours before death. So long as there is pain, some attempt is being made to repair the vital damages; but when pain ceases, then nature has given up the fight.

It is certainly a noteworthy fact that shock to the nervous system or the mind will induce a sort of stupor, and render pain absent, for the time being. Thus, Dr. Livingstone, the African traveller, relates that on one occasion he saw a lion which was just in the act of springing upon him:—

1 *Science and Immortality:* quoted by Dickinson, *Is Immortality Desirable?* p. 11.

2 *The Philosophy of Long Life,* pp. 225–6.

"He was on a little height. The animal caught him[1] by the shoulder as he sprang, and they both came to the ground together. Growling horribly close to his ear, he shook him as a terrier dog does a rat. The shock produced a stupor similar to that which seems to be felt by the mouse after the shake of the cat; it caused a sort of dreaminess in which there was no sense of pain nor feeling of terror, although he was quite conscious of all that was happening. It was like those experiences which patients partially under the influence of chloroform describe—who see all the operation, but feel not the knife. He claims that this condition was not the result of any mental process. The shake annihilated fear, and allowed no sense of horror on looking around at the beast. Fortunately he was rescued from his perilous condition without receiving any serious injury."

But, while it is generally asserted that the law of painless death is universal, unfortunately it must be admitted that there are some striking exceptions to this rule. Even very aged persons, who seem to view the approach of death with calm serenity, occasionally fight strenuously against it when the moment of final dissolution arrives, just as the convicted murderer, who knows that nothing can save him from the fate that is awaiting him in the person of the executioner, sometimes struggles so violently that the keepers find it necessary to drug him into a temporary state of semi-insensibility.

It is a remarkable fact that in certain cases the nearer the patient is to the point of death the more indifferent to it does he become. There are many instances on record in which the patient has fought against the oncoming of death for many hours or even days, but shortly before death occurred assumed a placid and

1 Related in the third person, and re-written from dictation.

peaceful expression and even attitude of mind. In some cases this is doubtless due to the accumulation within the system of carbon-dioxide and other toxic substances, which serve as deadeners to the sensitive nerves, and induce practical insensibility. But there are also cases on record in which the mind has remained apparently clear to the last, and yet no aversion to death has been manifested by the patient, though he was in terror of it before. Examples of this will be found in the next section.

3. The Consciousness of Dying

"Sir Benjamin Brodie states that he has been curious to watch the state of dying persons, and is satisfied that where an ordinary observer would not for an instant doubt that the individual is in a state of complete stupor the mind is often very active even at the very moment of death.

"Dr. Bailie once said that 'all his observations of death-beds inclined him to believe that nature intended that we should go out of the world as unconscious as we came into it.' 'In all my experience,' he added, 'I have not seen one instance in fifty to the contrary.' Yet even in such a large experience the occurrence of 'one instance in fifty to the contrary' would invalidate the assumption that such was the law of nature (or 'nature's intention,' which, if it means anything, means the same). The moment in which the spirit meets death is perhaps like the moment in which it is embraced by sleep. 'It never, I suppose' (says Mrs. Jameson, whose observations we quote), 'happened to any one to be conscious of the immediate transition from the waking to the sleeping state.' "[1]

A letter on this subject is to be found in the *Journal* of the (English) Society for Psychical Research, June 1898,

[1] *Mysteries of Life, Death and Futurity*, by Horace Welby, p. 147.

pp. 250–55, and I quote that part of it which bears upon the problem before us. It deals particularly with the subject of the consciousness of dying.

". . . From the materialistic point of view it would seem difficult, if not impossible, to account for such a phenomenon (as the consciousness of dying). Thus, if materialism be true, death must be the extinction of consciousness. It would seem that it must be impossible ever to be *conscious* of dying; that is, conscious that consciousness is being extinguished. Consequently, materialism would seem to make impossible the phenomenon which is at least an apparent fact. . . .

"I have stated the *a priori* difficulty in supposing the fact, and this is the circumstance that direct proof must be found in the experience of the individual himself who is dying, and external observers can only conjecture the condition of consciousness of the dying. But there is another difficulty. Often enough a person fears that he is dying when he is not, and also we often observe cases where persons evidently near death think that they are dying, when, in fact, they may survive hours, days, weeks, or even recover altogether. When, therefore, we measure such instances against those which happen to be connected with actual death, we may raise the question whether they are not after all merely inferences on the part of the decedent, and not immediate cognitions of it. Then, again, in favour of materialism and against the hypothetical assumption here made, we have to meet the allegation that we can be conscious of going to sleep, which on a materialistic theory ought to be as impossible as any alleged consciousness of dying, though the fact of going to sleep is perfectly consistent with materialism. Hence, if I can be conscious of going to sleep, which may be only a temporary suspension, as death is the permanent suspension of consciousness, why, the materialist will ask, may it not be possible to be conscious of dying? All these facts throw the burden of proof on the anti-materialist."

The writer (Dr. Hyslop) attempted to meet these arguments in several ways. First, he pointed out that many persons are never conscious of going to sleep. Yet one might be conscious of going to sleep without being conscious of dying. But it would appear, at all events, that a consciousness cannot be aware of its own suspension. It might be aware of its own *withdrawal,* but not its extinction; and the obvious inference to be drawn from this fact is that consciousness is probably withdrawn in both cases—sleep and death. This would agree with the traditional conception of the departure of the soul from the body. Certainly there seemed to be a consciousness, and a distinct consciousness, of dying in the case observed by him. And what is significant about the case is that his father (who was the patient observed) afterwards "communicated" through Mrs. Piper, apparently, and confirmed some of these inferences regarding the moment of death and the consciousness of dying! To be conscious of a thing we must possess a large amount of consciousness, and be able to reason clearly; and if consciousness were being extinguished at that time it would seem quite impossible for any person ever to be conscious of dying. The inability to express thought in motor action might be present, but that is a very different thing from an extinction of consciousness. Sometimes, indeed, there may be an intensely active consciousness, and yet it may be totally unable to express itself. In paralysis this is often the case; and when certain drugs are administered the body is unable to show any signs of consciousness, and yet all the senses and the mind are painfully active. It may be the same here. It is probable that at death there is a partial extinction of consciousness owing to the shock and wrench of death, and in the majority of cases this would doubtless prevent the individual spirit from exhibiting any external signs of consciousness; indeed, there was but little there—though we must always bear in mind the great distinction between the state of being conscious and the ability to express that consciousness in motor action. This is a distinction which is frequently overlooked by psychiatrists, but it should receive their careful attention. This subject of the consciousness of dying persons should certainly receive most careful attention from all physicians and others who have opportunities for studying the dying.[1] A tremendous mass of valuable psychological information might be gained in this manner, and it might throw light on the human spirit, its destiny and its potentialities, that could be obtained in no other way.

In one case known to me, a most interesting and suggestive phenomenon took place. The patient, who knew that she was dying, was dictating her last wishes—verbally—to those about her. Within a few minutes of her death she became too weak to speak, and requested that a pencil be placed in her right hand, and a pad of paper under the point of the pencil, so that she might write without hindrance. Her hand then proceeded to write out her dying wishes in a perfectly clear handwriting. The hand seemed to possess remarkable strength—a force of its own—the writing being bold and distinct, and the ideas conveyed were consistent and logical to the end. While this writing was going on, however, the patient completely lost control of her body; the breathing became stertorous, and she passed into a

[1] A number of such cases are to be found in a little book entitled *X-Rays*, by Gail Hamilton.

state of seeming unconsciousness. This state grew deeper and deeper, until the patient passed into a condition which might have been pronounced "death." The pulse and respiration ceased, to all appearances; the temperature fell; a limpness of the whole body ensued; the face became deathly pale, and yet her right hand and arm continued to write and write, and correct, and give clear and intelligible messages, which could only be interpreted as issuing from a sound and alert consciousness—in full possession of all its faculties. Where this intelligence resided I cannot say, but there can be no question as to its actual existence during the dramatic scene. The dead, inert body on the bed, the right hand and arm alive, mobile, active—writing out the behests of that consciousness—the whole scene came as closely as anything well could to a distinct utilization of a dead body by a living "spirit." It seems to me to bridge the gulf which separates normal, conscious influence from the automatic writing of an entranced medium.

Another fact of considerable interest in this connection, and bearing more or less directly upon the problem, is the question of *sensations* (mental operations) by those who *thought* they were dying, and who, as a matter of fact, revived, after they had lost consciousness.

In connection with this question of the existence of consciousness, and of its relation to the organism during the time of sleep, trance, etc., I desire briefly to refer to the argument by F. R. C. S. in his article, "Hora Mortis Nostrae," in the *Contemporary Review* for August, 1905. After emphasizing the fact that there is probably no pain at the moment of death in the majority of cases, and quoting Sir James Paget, who was even inclined to the opinion that if we were conscious of death it would be a pleasure—he states his own opinion—which is, that it is a condition involving neither pain nor pleasure, but is, on the contrary, a condition of total unconsciousness. In support of this he cites certain facts, observed by himself, of the effects of anaesthetics upon patients. Here he points out that after the senses have been obliterated one by one, there probably comes a moment when the patient is conscious of but one fact still left standing —that he is he. At that moment, if he be of a logical turn of mind, he may expect that he will now get behind the veil, see things as they are in themselves, contemplate pure Being, stand before the *merum ens* of his philosophy—and then somebody says, "He'll be all right now; it's a good thing he had it done"; and behold he is back in bed sick and sore, and drunk, painfully sorting unpleasant phenomena, and as far as ever from pure Being!

The writer turns to what he conceives to be the likeness between anaesthesia and death. He asks if we can find any clue to the nature of death in such states, and he is inclined to think that we can. His conclusion is in favour of materialism, as against the possible persistence of consciousness after death. He insists that these phenomena conclusively prove that no such thing as a soul-entity existing apart from the body is possible. He says:—

"To the notion of the soul as an invisible personage, made, and put into the body at birth, and extracted from it at the end of life, they [these facts] are utterly opposed. The anaesthetized body contains nothing save that which is bodily; no spark or vestige of consciousness; there it lies, still working, but without an occupant; just pumping the blood through the vessels, and maintaining the physical interchanges of the tissues; and if the loss of consciousness be due, not to an anaesthetic, but to injury, or disease of the brain, it may last an interminable time. Here, in these cases, is the best object lesson in materialism ever given to the world. . . . No amount of corpses can advance materialism, but to watch day after day a case of profound unconsciousness, the body a mere log, fed through a tube, fouling the bed, a physiological machine, a thing with no more thought in it than a dummy figure, and to see men and women brought to a like state in a few minutes by chloroform or ether, and kept there, just as part of the day's work; and to see the process reversed, and the lost owner of the body spirited back into it by an operation on his brain—here are the arguments ready made for materialism to be used with effect."

The writer sees no way out of this difficulty, since, as he says, if the mind be still there, in the anaesthetized body, with consciousness suspended, what is this mind, and where? To these questions he can find no answer.

Now it seems to me that a solution of these facts may be found, were we to conceive the relation of consciousness to organism from a different point of view than is afforded by present-day physiology, and the current "production" theory of consciousness. If the brain were the actual *producer* of consciousness, as is taught, of course its annihilation would be the only rational conclusion at which to arrive from these facts; but there is another way of viewing and interpreting these same phenomena. Consciousness might exist apart from the body, be *en rapport* with it, and merely manifest *through* it. On that theory the brain and nervous system, and even the body as a whole, would act merely as its transmitter, or vehicle for expression, and the paralysing of any centre in the brain by means of drugs, chemicals, etc., would mean simply that we have rendered impossible the motor expression of consciousness; we have rendered its manifestation to our sense perception impossible, but we have by no means proved that we have annihilated consciousness. We could take the same facts, and merely interpret them in a different manner. The conclusion which the author has drawn is therefore unwarrantable, and all his facts might be just as readily explained on the theory of an external consciousness or soul, which is active at the time elsewhere, and which persists after the death of the body.

Let me illustrate this point:—

Dr. Stephens (*Natural Salvation*, pp. 179–80) says:—

"What happens at death?

"First, the interlacing neurons let go their hold on each other, and self-consciousness of the person vanishes. It goes out, as flame vanishes when atoms of carbon and oxygen no longer combine.

"What next?

"The heart no longer propels the life-tide of refined food in the blood to the brain—as in sleep—and after a few minutes the neurons themselves die from suffocation and starvation. All those thousands of little individual lives vanish, as did the larger self-consciousness of the person; for in each the constituent bond of living molecules, atoms, and ions is disrupted.

"What next?

"The dissipation of the brain as cadaver is a somewhat slower, more homogeneous process, involving invasions of bacteria, disintegration, and reduction to more stable compounds, but tending ultimately to a return from the highly complex living substance, with all its maze of organization, to the abysmal base of the primeval ions and their lowly endowment of life-potential."

Here, it will be seen, we have as the first and most important condition the abolition of self-consciousness. It is considered the *sine qua non.* Yet we have seen that in many cases this self-consciousness is not abolished in the manner supposed at all; but that it is conscious and aware of all that is going on. As I have argued, this does not look in the least like extinction, but rather transition. And again, it would be most difficult to account for many of those cases in which the subject had dropped dead *instanter.* Are we to suppose that the interlacing neurons let go their hold on each other all at once? Or would it not rather appear to be an instantaneous process complete in itself, and that this "letting go" phenomenon was merely one of the many physiological processes that *resulted* from death, rather than the one that *caused* it? We must be most careful to distinguish between cause and effect here. Does consciousness cease because the neurons no longer function; or do the neurons cease to function because consciousness is no longer present? Of course that is always ground for debate, and is a question that has not yet been settled—opposite schools taking opposite views—and although I cannot claim that the facts tell in favour of my theory, yet I must insist that they do not tell in favour of the opposing theory either.

And there are certain analogous facts which would seem to indicate that the cessation of mental activity is the primary and most important phenomenon—the nervous activity following it. Thus, Professor Mosso has shown that when a subject awakens from sleep, mental operations begin at once, and the increased flow of blood to the brain follows as a secondary phenomenon. Here it is not the afflux of blood which insures the mental activity, but the greater activity of consciousness which necessitates the increased flow of blood to the brain. As William James says in his *Principles of Psychology* (Vol. I, p. 99):—

"Many popular writers talk as if it were the other way about, and as if mental activity were due to the afflux of blood. But as Professor H. N. Martin has well said, 'That belief has no physiological foundation whatever; it is even directly opposed to all that we know of cell-life.'"

It will be seen, therefore, that these facts point just as strongly in one direction as in the other. Decisive "proof" *pro* or *con* is still lacking.

4. Sensations While Falling

Remarkable flashes of memory have frequently been recorded, while the subject was *drowning.* There are a few cases on record in which similar mental flashes have been observed by persons falling great distances, and it is probable that we should have a larger number of such examples if more people who had fallen great distances had lived to tell the tale. The following cases, however, cannot fail to be of interest in this connection.

The following is a typical example of a case of this character. The psychological interest is remarkable. It runs, in part, as follows:—

"Although I fell backward from a tremendous height, I experienced none of the anxiety which occasionally attacks us in dreams at supposed falling accidents; on the contrary, I felt as if I were carried downwards slowly on giant wings that protected me against collision. During the whole time of this fall, consciousness never left me. Without feeling the least bit embarrassed or frightened, I reviewed my situation and the future of my family; and the various features of my own life passed before me with unequalled rapidity. I have heard people say that, in falling a great distance, one loses his breath; I never lost my breath, and when my body finally bounded against the rocks at the foot of the glacier, I became unconscious without experiencing any pain whatever. I felt nothing of the many wounds on head or limbs received during my journey down the precipice from coming in contact with rocks and masses of ice. The moments when I stood at the brink of a future life were the happiest I ever experienced. I remember reading the provisions of my life insurance policy with my mind's eye; the big sum of money which death was bound to bring to my loved ones I saw before me counted out on a green table-cloth, all in crisp bills and shining gold."

Dr. Heim gives the following description of his fall down a mountain side, which he fully expected would end in certain death:—

"Quick as the wind I flew against the rocks to my left, rebounded, and was thrown upon my back, head downward. Suddenly I felt myself carried through the air for at least a hundred feet, to finally land against a high snow wall. At the instant I fell, it became evident to me that I was to be thrown against the rock, and I did my utmost to avoid that calamity by digging with my fingers in the snow and tearing the tips of them horribly without knowing it. I heard distinctly the dull noise produced when my head and back struck against the different corners of the rock; I also heard the sound it gave when my body bounded against the snow wall, but in all this I felt no pain; pain only manifested itself at the end of an hour or so." (*Encyclopedia of Death,* vol. ii, pp. 384–5).

5. Memory at the Point of Death

These apparently supernormal flashes of memory and conscious activity at the moment of death are of very great interest from many points of view. If memory be a purely physiological process, as materialistic psychology would have us believe, how comes it that such instantaneous and vast recallings are possible—at a time too, when the brain is supposed to be in a *lessened* condition of activity? It is not that the brain is preternaturally stimulated at such times, precisely the reverse; it is practically *inert* and unresponsive to external stimuli; and, one would think, would be in no condition to think and remember normally, far less recall such immense numbers of facts in so short a period of time. And not only is the time remarkably brief on such occasions, but facts are often recalled which had entirely passed out of the conscious mind, and would never have been remembered in the normal course of the conscious life. It would almost seem that *nothing is forgotten*—a statement which agrees with De Quincey's estimate of the case. In his *Opium-Eater,* he says:—

"Of this, at least, I feel assured, that *there is no such thing as forgetting possible* to the mind; a thousand incidents may and will interpose a veil between our present consciousness and the secret inscriptions of the mind; accidents of the same sort will also rend away this veil; but alike, whether veiled or unveiled, the inscription remains for ever, just as the stars seem to withdraw before the common light of day, whereas, in fact, we all know that it is the light which is drawn over them as a veil, and that they are waiting to be revealed when the obscuring daylight shall have withdrawn."

Similarly, the author of the *Hasheesh Eater* says:—

"De Quincey's comparison of it to the palimpsest manuscripts, while it is one of the most powerful that even that great genius could have conceived, is not at all too much so to express the truth. We pass, in dreamy musing, through a grassy field; a blade of the tender herbage brushes against the foot; its impression hardly comes into consciousness; on earth, it is never remembered again. But not even that slight sensation is utterly lost. The pressure of the body dulls the soul to its perception; other external experiences supplant it, but when the time of the final awakening comes, the resurrection of the soul from its charnel of the body, the analytic finger of inevitable light shall search out that old impression, and to the spiritual eye, no deep-graven record of its earthly triumphs shall be clearer!"

Surely this closely resembles the *Book of Judgment* of Theology!

CHAPTER III

THE CAUSES OF DEATH

1. Sudden Death

In the present chapter I propose to give a brief résumé of the main (known) causes of sudden death. Our chief interest is the study of natural, and not unnatural, death—as all *sudden* deaths are. When death results from any disease, it is tolerably clear to us what the actual cause of the death is, in that case. We can at all events form a mental picture, in rough outline, of what has taken place; but the same is not true in cases of sudden and unexpected death. Often the cause is most difficult to find, and it must be acknowledged that, even here, much is still uncertain and unknown. Much less is known of the nature and cause of "natural" death—as we have seen, and shall see further. Before we proceed to a consideration of this last and most vital question, however, we must first of all consider sudden death, arising from various causes—when such causes are known.

In the first place, then, it may be said that such a thing as "sudden death" does not, strictly speaking, exist at all! In those cases where it is supposed to have taken place, it can almost invariably be shown that some cause or causes, acting for considerable periods of time upon the body, have produced these results. Says Dr. Brouardel:—

"Why does sudden death occur? No one dies suddenly, apart from the effects of violence, so long as all the organs are sound; but there are some diseases which develop slowly and secretly, without the attention of the patients having been called to them by any pain or by any feeling of illness, and without a physician ever having been called in, and which terminate naturally by a rapid death. . . . We will define sudden death as 'the rapid and unforeseen termination of an acute or chronic disease, which has in most cases developed in a latent manner.' . . . However carefully we may perform every autopsy, however minute our exploration of the body may be, however thorough may be our knowledge of the causes of death, we sometimes meet with cases which it is impossible to explain. The proportion is about 8 or 10 per cent."

This is a very significant admission, of which I shall have occasion to remind the reader at a later stage of our investigation.

Turning now to the causes of natural death, we find the first place occupied by lesions of the heart and circulatory system.[1] And we read that "a lesion may remain latent during the greater part of life, and be only revealed by accident" (p. 74). Lesions of the heart may result from a number of causes—(1) fatty overgrowth of the heart; (2) fatty degeneration of the mus-

[1] If an artery breaks, that is said to be the cause of the death of the individual, but few stop to ask, "Why should the artery break?" Would it not be more accurate, strictly speaking, to say that the real cause of the person's death was *that* cause which so weakened the wall of the artery that its rupture was possible? Or, if death takes place owing to some central inhibition, would it not be better to seek the cause of the inhibition rather than rest content with the mere verdict of "heart failure"? To all thinking persons, the true causes of death lie deeper than the mere effect or resultant—the "last straw that broke the camel's back" in very truth! Strictly speaking, the cause of death, in such cases, is *the cause of this last cause;* and, what *that* is, I have tried to show in another place.—*Vitality, Fasting and Nutrition* (pp. 324–31).

cular tissue of the heart; (3) fibroid degeneration of the heart; (4) lesions of the coronary arteries; (5) syphilitic affections of the heart; (6) rupture of the heart, etc. Then we have lesions of the pericardium. Following this, as causes of sudden death, we have mitral and tricuspid incompetence, endocarditis, *angina pectoris,* and neoplasms of the heart.

In looking up the literature on death I came across a rare manuscript, viz., lectures on medical jurisprudence given by Sir Douglas Maclagan in 1888. This manuscript is in pen and ink, and is doubtless the only one of its kind in existence. Amongst a variety of interesting topics, the subject of death in its various forms is treated by the author at great length, and certain facts are detailed in this manuscript that I have not discovered in any book printed and on the market. There are, however, certain statements contained in the book with which I can by no means agree. Thus, our author includes under "natural death," deaths due to hemorrhage, diarrhoea, wasting diseases, deficient power, organic lesions, apoplexy, toxaemia, epilepsy, mental emotion, perforation of the viscera, closure of the glottis, congestion of the lungs, effusion in the lungs, diseases of the spine, paralysis, and tetanic spasm. With the single exception of deficient power, I should hesitate to class any of the above deaths as natural. Deaths due to disease are invariably *un*natural and premature. It is for this reason that I have not included in this volume deaths due to disease, murder, suicide, and infanticide. (It may be well to state in this place also, that I have omitted all discussion of death from the medico-legal point of view, this seeming to me out of place in a work of this character.)

We next come to lesions of the arteries. Here we find: Congenital lesions, arteriosclerosis, aneurisms, spontaneous rupture of the aorta. Of the veins: various ruptures, also air in the veins. There are also lesions of the capillaries, miliary aneurisms, meningeal haemorrhages, capillary embolisms, local disturbances of the circulation. Some of these, it may be said, can hardly be classed as causes of sudden death, in the orthodox meaning of that term.

A large number of sudden deaths are due to lesions of the cerebro-spinal system and the major neuroses. Here we may classify meningitis—tubercular, chronic, cerebro-spinal, etc. Abscesses of the brain, cerebral tumours, lesions of the spinal cord, lesions of the nerves, epilepsy, hysteria, inhibition, and sudden death from emotion or mental causes. I shall have occasion to recur to this latter cause of death, when we come to consider its character in cases of "natural death."

There is, next, a whole set of causes of sudden death due to lesions of the respiratory system. Among these we find: lesions of the larynx, of the trachea, of the thyroid body, of the mediastinum, pulmonary congestion, pneumonia, capillary bronchitis, pulmonary phthisis, cancer of the lung, emphysema of the lungs, pleurisy, rupture of the diaphragm, and compression of the chest.

Next, we have lesions of the digestive system. These are:—Lesions of the pharynx, of the oesophagus, of the stomach (which might include a number of subdivisions), lesions of the liver, of the spleen, of the pancreas, and of the suprarenal capsules. Among doubtful causes (to me) are included corpulency, climatic excesses of heat, cold, etc. These can hardly be called *causes* of sudden death;

rather, *occasions* of sudden death—when the organism is already in such a state that life can easily be terminated by a very slight mal-adjustment of external circumstances.

In the female there are also special causes of death, to which the male is not subject. Among these are: vaginal examination, extra uterine gestation, recto-uterine haematocele, rupture of the uterus, vulvo-vaginal varices, syncope arising out of uterine conditions, etc.

There are also many cases of sudden death in fevers and kindred states—in anthrax, mumps, diphtheria, acute rheumatism, typhoid fever, plague, etc., etc. A very full study of death from some of these conditions will be found in Dr. John D. Malcolm's *Physiology of Death from Traumatic Fever* (London, 1893). Sudden death may also be due to haemophilia.

Sudden death may take place in various diseases which cannot of themselves be said to be the cause of the death—e. g., in diabetes, uraemia, gout, dropsy, as well as in cases of alcoholism. In children, sudden death may result from syncope, convulsions, asphyxia, pulmonary congestion and various intestinal disorders. All of these classes are subject to various subdivisions. They will all be found discussed in full in Dr. Brouardel's book on *Death and Sudden Death*, to which excellent manual I would refer the reader for further particulars regarding such cases.

Death from Burns and Scalds.—A great intensity of heat is not required to destroy vitality of the skin. The danger to life is much more in proportion to the extent of surface of the body exposed to the action of the fire than to its intensity. Sometimes it may prove fatal by setting up inflammation of the internal tissues. In the case of corrosions with acids, the marks are generally of a dirty brown colour.

Death by Haemorrhage.—The body is blanched. On dissection, the great venous trunks are flabby and empty. The large regions internally are pale. We may find hypostasis in the inferior parts of the lungs, even when death is caused by haemorrhage. We may find evidence of haemorrhage in the internal parts, generally partly fluid and partly clotty. It is often quite impossible to detect from what vessel the blood has come.

Dr. Harrison, writing years ago on death from haemorrhage, said, in his *Medical Aspects of Death:*—

"Death may be said to begin at different parts of the body; and it will be found that the nature, symptoms, and peculiarities of the act of dying are determined by the organ first mortally attacked. The alterations which directly occasion dissolution seem principally effective either in the arrest of the circulation or the respiration.

"As the heart is the great mover in the circulation, we can easily conceive that whatever brings it to a stop must be fatal to life. Extensive losses of blood operate in this manner, and they furnish us with a good illustration of the manner in which death takes place. The sufferer becomes pale and faint, his lips white and trembling; after a while the breathing becomes distressed, and a rushing noise seems to fill the ears. The pulse is soft, feeble, and wavering; the exhaustion and prostration are more and more alarming. Soon a curious restlessness takes place, and he tosses from side to side. At length, the pulse becomes uncertain, and the blood is feebly thrown to the brain. The surface assumes an icy coldness; the mind is yet untouched, and the sufferer knows himself to be dying; in vain the pulse is sought at the wrist—in vain efforts are made to re-excite warmth—the body is like a living corpse. Now, a few convulsive gaspings arise, and the countenance sets in the stiff image of death. Such are the more striking phenomena which attend the fatal haemorrhages.

"The failure of the vital powers, from the withdrawal of blood, may be regarded as a sort of type of this mode of death, since the various symptoms which have been named arise from the cessation of the healthy circulation.

"A dread of the loss of blood may almost be considered as an instinctive feeling; at any rate, its importance is early impressed on the mind, and is never forgotten. In childhood, it is looked at with alarm; and the stoutest mind cannot but view with horror those perilous gushes of blood which bring us into the very jaws of destruction."

2. Mental Causes of Death

Disease and death are more frequently the effect of mental causes than might be supposed. Whenever there exists a predisposition to apoplexy, close mental application is always attended with the utmost danger, especially in the latter part of life. Epilepsy is another disease of the nervous system that may be induced, or exaggerated, by the state of the mind, and extreme mental dejection, hypochondria, and even insanity, may sometimes result from these causes. Many individuals distinguished for their special talents and learning have been subject to such unhappy maladies, and yet it is difficult to determine how much of the disease may justly be ascribed to the abstract labours of intellect, and how much to mental anxiety, for it is known that undue strain upon the emotions—either excitement or depression—may be productive of these results.

Thus, in the case of Sir Walter Scott, the extreme literary labours that he performed do not seem to have had any injurious effect upon his health, until the brightness of his fortune had become overcast by the clouds of adversity. When, with his mental tasks, were mingled the agitating emotions of anxiety, resulting in irregularity of habits, his physical health began to break, and the fatal disease of the brain soon brought a tragic ending to his life.

While there may be occasions when even the ordinary exertions of the brain are attended with danger, their effect upon the health is usually comparatively slight, unless they are combined with one or more of the numerous feelings, pleasurable and painful, which, according as they are mild or intense, are known to us as emotions or passions.

As Dr. William Sweetser said, in his work, *Mental Hygiene:*—

"The agency of the passions in the production of disease, especially in the advanced stages of civilization, when men's relations are intimate, and their interests clash, and their nervous susceptibilities are exalted, can scarcely be adequately appreciated. It is doubtless to this more intense and multiplied action of the passions, in union, at times, with the abuse of the intellectual powers, that we are mainly to attribute the greater frequency of the diseases of the heart and brain in the cultivated than in the ruder states of society. Few, probably, ever suspect the amount of bodily infirmity and disease resulting from moral causes—how often the frame wastes, and premature decay comes on, under the corroding influence of some painful passion. . . . In delicate and sensitive constitutions, the operation of the painful passions is ever attended with the utmost danger; and should there exist a predisposition to any particular form of disease, as consumption, or insanity, it will generally be called into action under their strong and continued influence."

While it is undoubtedly true that some passions act most obviously upon the heart, others on the respiration, and some on the digestive organs, it has been clearly proved that, so far from being limited to one particular organ, a number of the organic viscera are almost invariably included within the influence of a strong emotion. At the same time, such a close correspondence exists between the mental or moral feelings and the physical body, that the condition of the former may either determine or be determined by that of the latter. For example, indigestion may sometimes be the cause, and sometimes the consequence of an irritable or unhappy temper. "Sour stomach" may either occasion or result from a sour disposition. To sweeten one is certain to have a neutralizing effect upon the other. It is, therefore, obvious that an unhealthy mental state imparts an unhealthy influence to the bodily organism, and, if such evidence were needed, scores of historical facts might be cited to establish the truth of this fact.

The pleasurable emotions—love, hope, friendship, pride, joy, etc.—may, if properly experienced, produce an expansion of vital action, and yet even these emotions, if felt intemperately, exert a very contrary effect. The expression that "joy kills" has a basis in fact, for, as Haller says, in his *Psychology*, "Excessive and sudden joy often kills, by increasing the motion of the blood, and exciting a true apoplexy." It is said that Pope Leo X died from the effect of extravagant joy at the triumph of his party against the French; and Dr. Good, in his *Study of Medicine*, cites the case of a clergyman who, at a time when his income was very limited, received the unexpected tidings that some property had been bequeathed to him. "He arrived in London in great agitation; and, entering his own door, dropped down in a fit of apoplexy, from which he never entirely recovered."

If such facts are true in regard to the pleasurable passions, there is much more danger of injurious results when the emotions are of a painful character. To quote Dr. Sweetser again:—

> "The painful passions act immediately upon the nervous system, directly depressing, disordering, expanding, and sometimes even annihilating its energies. . . . Although the general effect of the painful emotions is to induce a contraction or concentration, and a depression of the actions of life, yet, in their exaggerated forms, they are sometimes followed by a transient excitement, reaction, or vital expansion, when their operation, becoming more diffused, is necessarily weakened in relation to any individual organ. Under such circumstances, the oppression of the heart and lungs is in a measure removed, and the circulation and respiration go on with more freedom. Hence it is that when anger and grief explode . . . their consequences are much less to be dreaded than when they are deep, still, and speechless, since here their force is most concentrated."

Thus, in extreme paroxysms of anger, the physical phenomena are most apparent. The face becomes distorted and repulsive, the eyes sparkling with brutal fury. All the vital actions are oppressed, and often are nearly overwhelmed. The blood retreats from the surface; tremors and agitations appear in the limbs, or perhaps in the entire body, and there is frequently indication of excessive nervous affections, sometimes giving place to sobbing and hysteria, and sometimes to convulsions and spasms. The action of the heart is also affected, becoming feeble, laboured, irregular. and even painful. The effect upon the respiration is shown in the short, rapid, and difficult breathing, which produces a feeling of suffocation, a tightness that is felt in the whole chest, and that occasionally extends to the throat, choking, and otherwise interfering with the power of speech. If not noticed at the moment of anger, the influence of this passion almost invariably proceeds to the abdomen, as indicated by the subsequent distress appearing in the region of the stomach, this being due to the disturbance of the stomach, liver, and bowels.

Almost innumerable instances are known in which fainting has resulted from violent anger, and in many cases life itself has paid the price of this paroxysm of the emotions. According to John Hunter, the eminent physiologist, death from anger is as absolute as that caused by lightning. In such cases, the muscles remain flaccid and the body passes rapidly into putrefaction.

Dr. Hunter himself is one of the historical victims of anger. Though a man of extraordinary genius—as all medical men know—he was subject to violent passions which he was never able to control. When engaged one day in an unpleasant altercation with his colleagues, some of whom had peremptorily contradicted him, he became too angry to continue speaking, and, hurrying into an adjoining room, instantly fell dead. The direct cause of his death was, of course, the affection of the heart from which he had long been a sufferer, but there can be no question but that the final stroke was superinduced by anger.

Tourtelle, the French physician, asserted that he had "seen two women perish--one in convulsions at the end of six hours, and the other suffocated in two days—from giving themselves up to transports of fury."

Anger destroys the appetite and interferes with the functions of digestion, and Dr. Beaumont, who was first able to look into the human stomach through the opening caused by a fistula, discovered that anger or other severe mental emotion, "would sometimes cause its inner, or mucous, coat to become morbidly red, dry, and irritable, occasioning at the same time a temporary fit of indigestion."

The unpleasant dryness of the throat caused by anger —a condition which occasions the frequent swallowing action of the muscles—is due to the inspissation of the saliva; and some authorities have even gone so far as to assert that such an exhibition of emotion may cause the fluid of the mouth to acquire poisonous qualities "capable of provoking convulsions, and even madness, in those bitten by a person so agitated." [1]

It is well known that haemorrhages from various parts of the body—the nose, lungs, and stomach—as well as inflammations of different organs, may be produced by severe attacks of anger; and Dr. Sweetser asserts that he himself has "now and then met with instances of erysipelatous inflammation about the face and neck, induced by paroxysms of passion."

Irritability and moroseness of temper, when long continued, may also cause inflammatory and nervous disorders, and it is well known to physicians and surgeons that the fretful and fractious patient recovers less promptly, and is more exposed to relapses, than he who is possessed with a quiet resignation to existing conditions. Wounds that have healed have even been known to break out afresh as the effect of unfavourable mental conditions.

1 Broussais' *Psychology*.

Fear, like anger, has its degrees; and its effect upon the health depends upon its intensity. When extreme, however, the results are often astonishing. Thus, in acute fear, the respiration becomes immediately and most strikingly affected. At the first impulse, a sudden inspiration occurs, owing to a spasmodic contraction of the diaphragm, and this is immediately followed by an incomplete respiration, cut short apparently by an internal spasm—either of the throat, windpipe, or lungs. The effect upon the respiration is to make the breath short, rapid, and tremulous. The voice trembles, and, because of the diminution of secretions of the mouth and throat, becomes thick and unnatural. At times, even speechlessness may follow.

Naturally the heart suffers from the effect of such an acute sensation. Being oppressed, or constricted, it flutters or palpitates, and in other respects is visibly agitated. Consequently, the pulse also becomes irregular. The viscera of the abdomen experience disagreeable effects from the sensation of fear, and these frequently show themselves in spasmodic contractions, or in a morbid increase of secretions. Occasionally vomiting, but more frequently a somewhat involuntary diarrhoea, occurs. The urine also, though increased in quantity, becomes pale and limpid, and there is an urgent if not absolutely irresistible desire to void it frequently. These latter symptoms, it may be added, are frequently shown in other forms of serious attack upon the nerve force.

In time of fear, the blood leaves the surface so perceptibly that the face becomes pallid, while the skin, sometimes in all parts of the body, grows cold and rough, or, as we commonly say, like "goose-flesh." Frequently this apparent chill breaks forth in a cold sweat on the forehead, and often in other parts of the body as well. Even the hair of the head may become elevated, and the general tremor or shuddering, that attacks the limbs, proceeds to the teeth, producing a chattering sound very similar to that which is exhibited under conditions of extreme cold, or in a paroxysm of fever. As in the case of anger, fear may induce most painful and unnatural contortions of the countenance, with convulsive sobbing and, in the case of women especially, tears; or, under extremely violent emotions, hysteria. Even in men, however, the depressing effects of fear sometimes include the entire chest and upper part of the abdomen within their field of influence, and if the sense of constriction becomes too agonizing, syncope and sometimes death itself may follow. Just as a sudden though brief attack of anger may arrest digestion and disarrange the entire nervous organism for a whole day, so fear exerts a most dangerous effect upon the nerves and muscles, sometimes even acting as a sudden cathartic.

If the expression, "frightened to death," is no idle jest therefore—and there is no lack of examples to prove that hundreds of persons have been literally frightened out of existence—this fear, when severe, but less pronounced, may exert a distinctly contrary effect. Thus, while convulsions, epilepsies, and even insanity, have resulted from this emotion, these, as well as many other affections, have been immediately suspended or entirely removed, by a strong expression of this feeling. It sometimes surprises us to note how quickly a toothache stops when we enter the dentist's rooms, but we seldom analyse the mental process carefully enough to determine that it is the fear of the greater pain of extraction that makes the minor nervous affection less. What is true in regard to the toothache also applies to many other ills, including sea-sickness, hypochondria, etc.

The horror which we feel in the presence of insects, reptiles, and other creatures known to be entirely harmless, is but another form of fear, and its effect upon the physical organism is almost as distinctly pronounced. Thus, there is the same sudden paleness and coldness; the contraction of the skin and elevation of the hair; the chills and rigours of the body; the panting and oppression of heart and lungs. When greatly aggravated, the conditions of deadly fear—the convulsions, epilepsy, and even instant death—are realized. Thus, Broussais refers to the case of a woman who, on feeling a living frog that had been dropped into the bosom of her dress, was seized with profuse bleeding from the lungs, and survived but a few minutes.

Such antipathies may be innate, like the terror that so many individuals feel at the sight of mice; and yet grown persons as well as children have been thrown into convulsions, and have even derived serious nervous injury, by being subjected to the immediate influence of objects that have been a source of repugnance or horror to them.

Grief, whatever its cause, is essentially a mental pain; and it is inevitably productive of physical phenomena. In its simplest forms, or when produced by the loss of kindred, friends, property, or other things that are generally deemed desirable, it is usually subdued by the healing balm of time; but when, as often happens, it is complicated with some one of the malignant emotions of the heart—hatred, revenge, jealousy, etc.—the mental pain is accentuated, and the deleterious effect is increased. As we cannot escape this suffering when we give way to the sentiments of envy or revenge, we punish ourselves by our hatred far more than we injure the object of these vicious feelings.

When grief is acute, it is usually transient in character. When it becomes chronic, it develops into melancholia. In its acute stage its symptoms somewhat resemble those of anger, for all passions founded on pain are closely related as to their effect upon the bodily functions. For example, there is the same agonizing feeling of impending suffocation; the sense of oppression and stricture at the heart and lungs. The entire chest feels as though tightly bound, and the demand for air to alleviate this oppression is indicated by the long-drawn or protracted inspirations. The greatest distress, however, is experienced in the heart, and, in moments of thrilling distress, this heart-agony becomes so great, that it is not uncommon for its victims to die—broken-hearted.

As in cases of anger, or fear, the influence of the emotion of grief also extends to the throat and mouth; it affects the circulation, weakening the pulse perceptibly, and, finally, proceeds to the organs of the abdomen, being experienced especially in the pit of the stomach. The appetite fails; the powers of digestion are impaired, or suspended, and the throat becomes so contracted that it is impossible for the victim to swallow food without frequent draughts of liquor to "wash down" every mouthful.

Those exhibitions of bodily anguish known as "sobbing," or "crying," represent one of the greatest safeguards in moments of grief. Thus, death from grief is said to be unknown in cases where the sorrow has been attended by copious weeping, for the tears relieve the oppression of the head and lungs, forming a sort of natural crisis to the paroxysm, just as sweating is the crisis to the paroxysm of fever.

Insanity and monomania, as well as many other nervous affections, not uncommonly follow in the wake of grief, just as they attend upon the emotions that we may term anxiety. In other words, worry also kills through its continued depressing effect upon the heart and other vital functions. Palsy, chronic inflammation, dyspepsia, are some of the various ills that may be induced by the protracted operation of the sentiments known as sorrow, anxiety, or worry, and from any of these disorders man may die.

In his *Philosophy of Long Life,* M. Jean Finot devotes a number of pages to a consideration of the problem: "Will as a means of prolonging life." He says in part:—

"The forces of the mind, well utilized, may render us most important services from the point of view of the prolongation of life, as we have demonstrated elsewhere. It is suggestion ill-employed which undoubtedly shortens it. Arrived at a certain age, we drug ourselves with the idea of the approaching end. We lose faith in our powers, and they abandon us. Under the pretext of the weight of age upon our shoulders we take on sedentary habits. We cease to busy ourselves actively with our occupations. Little by little our blood, vitiated by idleness, together with our ill-renewed tissues, opens the door to all kinds of diseases. Premature old age attacks us, and we succumb sooner than we need in consequence of a harmful auto-suggestion.

"Let us try to live by auto-suggestion instead of dying by it. . . . Evil suggestions surround us on all sides. . . . Just as the hypochondriac begins to beam with happiness by continually repeating that he is gay, so persons obsessed by the thought of old age and death will think calmly of their approach. The unreasoning fear of them, by demoralizing their consciousness, only quickens their destroying march. Man, arrived at a certain age, or even at a certain mental state, undergoes a sort of auto-suggestion of death. He then believes himself to have reached the end of his days, and feeds as much on the fear of death as on daily foods. From this moment onwards death fascinates him. He hears its call with terror everywhere and always. The philosophic and salutary consciousness of a hereafter gives place to a cowardly and nervous fear of separation from life. The victim feeds upon this fear, intoxicates himself with it, and dies of it!"

3. Death by Poisoning

It is customary to classify poisons as irritant, corrosive, or neurotic, according to their effect upon the system. At the same time, certain poisons are so complicated in their action upon the human organism, that one seems to present the characteristics of another. Thus there are some irritant poisons that exert a corrosive effect, although many do not, and, under certain conditions, every corrosive may act as an irritant.

Most irritant poisons belong to the mineral kingdom—being both metallic and non-metallic—although the vegetable kingdom supplies a few, while some of the gases also come within the province of irritants. Neurotic poisons, according to Taylor, "act upon the nervous system. Either immediately, or some time after, the poison has been swallowed, the patient suffers from headache, giddiness, numbness, paralysis, stupor, and, in some instances, convulsions." "But," as Griffiths says (*Police and Crime,* vol. ii., 159–60)—

"The symptoms of all kinds of poisons intermingle, and the irritants may produce the same as the neurotics, and some—those especially which are derived from the vegetable kingdom—have a compound action. But one and all are defined in legal medicine as substances which, when absorbed into the blood, are capable of seriously affecting health or of destroying life."

To again quote the same authority:—

"First among the irritants we may take sulphuric acid, or oil of vitriol, a poison often used in suicide, and in the form of vitriol-throwing to do injury without actually causing death. Nitric acid is the *aqua fortis* of the Middle Ages, often mentioned in the annals of poisoning. With nitric acid may be classed hydrochloric or muriatic acid, which was given by a servant at Taunton to her mistress in beer. Oxalic acid is a vegetable acid, generally very rapid in its action, and leaving, as a rule, little trace. Tartaric acid and acetic acid, although irritants in large quantities, are not commonly classed with poisons."

Cases of poisoning by phosphorus, an irritant poison, have been known for long in England, but are more common in France, the substance having generally been obtained from the tips of common lucifer matches. A girl at Norwich put some compound of phosphorus used for vermin-killing into the family teapot with murderous intent, but when hot water was poured upon the leaves the smell betrayed the poison. A woman put some phosphorus into soup she gave her husband, who began to eat it in the dark, when the luminosity of the liquid showed something was wrong.

Arsenic is the best known of the metallic irritants. There are so many preparations of it, that it is easily obtained; it is not difficult to give, for it imparts no particular flavour to food. The symptoms vary; they are shown within eight hours, but sometimes not for five or six days. This poison may be administered in small quantities, and spread over some length of time, so as to constitute chronic poisoning.

Arsenic is sometimes called "the fool's poison," because it so generally betrays its presence in the human body, even after long periods have elapsed. The body of Alice Hewitt—poisoned by her daughter—was exhumed after eleven weeks, and 154 grains of solid arsenic were found in her intestines alone. Other still more remarkable cases are recorded—one in which the poison was found in children after eight years' burial; a second case is quoted after twelve years had elapsed, and a third fourteen years. Arsenic has also the inconvenient action (from the murderer's point of view) of preserving the body and resisting decomposition. This has been exhibited for months, nay, years, after interment. It was seen to a marvellous degree in the case of Pel's wife, and in the Guestling poisoning. And yet again in St. Celens (France), where ten bodies were exhumed and found well preserved. Zinc chloride is another powerful preservative: it retards putrefaction by combining with the tissues. Palmer's wife was exhumed after twelve months' burial, and all organs had been preserved by the antimony with which she had been poisoned. Chloride of lime had the same effect in the case of Harriet Lane.[1]

The facility with which arsenic or some of its compounds can be purchased has no doubt multiplied its felonious use: this, and the plausible excuse so generally

1 For details of quoted cases see Griffiths, *Police and Crime*, 1899.

put forward when buying it, that it is to kill rats and other vermin—an excuse as old as Chaucer. Lady Fowlis, when indicted for witchcraft and poisoning in 1590, was accused of giving "eight shillings money to a person for buying rateoun poison."

Tartar emetic is a substance with an evil reputation in the chronicles of poisoning. Two famous cases are on record, although both are mysteries to this day—surrounded with such strong doubts that they should, perhaps, be removed from the records of crime.

Dr. Smith Ely Jelliffe, writing in the *Encyclopedia Americana,* admits that a strictly scientific definition of the word "poison" cannot be given.

"In general it is said," he adds, "that a poison is any substance which brings about a change in the molecular composition of an organ, or organs, causing its functions to depart very distinctly from the normal. But what grade of molecular disturbance is necessary to make a substance a poison, and how far from the normal must be the functional alteration, it is impossible to say.

"It is believed that for practically all forms of poison a distinct alteration in the character of the cells of the body takes place, as well as a change in the chemical composition of the poisonous substance. . . . It is rarely that the reaction between the body-cell and the poison is purely of a physical nature, yet this very frequently happens in many poisons that act on the blood. By some of the poisons—the anilines, for example—the blood undergoes changes, not so much due to new chemical compounds formed as in the physical changes in the tension of the blood serum and the blood corpuscles, whereby the blood-colouring matters stream out into the plasma, and the oxygen-carrying function of the blood is lost. Similar types of poisoning result from some of the metals, and the poison of the cholera organism is thought to act in a like manner. In other poisons there is a direct union of the ions of the poison with some constituents in the cells of the body, making new chemical compounds, and thus interfering with the molecular activities of the cells."

The following is a summary of some of the most common types of poisoning:—

Poisoning by the mineral acids—nitric, sulphuric, hydrochloric—is not uncommon. In these there is a marked caustic action, with intense burning pain when taken by the mouth. The lips are stained yellow, black, or white respectively, according to the poison taken. There is nausea, vomiting, and diarrhoea, with all the symptoms of an intense gastro-enteritis, with collapse, pale face, cold sweating extremities, small, feeble pulse, rapid respiration; and the patient dies in intense agony.

Poisoning by alkalies is infrequent. Occasionally sodium hydrate, or potassium hydrate, is swallowed. Lime is also taken by accident; so (rarely) is ammonia. The symptoms are much like those of poisoning by the mineral acids, except that there are no marked discolorations. The halogen compounds are very markedly poisonous as gases, notably chlorine, bromine, fluorine; and the iodides and bromides cause forms of chronic poisoning.

The heavy metals as such are not poisonous, but their soluble compounds are all poisonous. They vary widely, however, in strength. In order, from the strongest to the weakest, they are caustic or astringent. In all the symptoms are analogous; there is severe gastro-enteritis, with symptoms of collapse. According to the solubility or insolubility of the poison, the burning is more or less deep.

Arsenic and phosphorus are poisons that give very

similar symptoms: acute gastro-enteritis, with nausea, vomiting, purging; then some grade of apparent recovery, to be followed after a few days with a recrudescence of the gastro-enteritis and the development of secondary blood-vessel changes, which may cause minute haemorrhages in any part of the body. Then follow fatty degeneration and death.

Practically all of the anaesthetics and hypnotics belong to the alcohol group, and produce allied symptoms.

Phenols form a distinct group, in which carbolic acid may be taken as a type. This causes gastro-enteritis, with severe pain, white scars on lips and throat, buzzing, dizziness, smoky to blackish urine, pale, bluish face, weak heart, quick breathing, coma, and sometimes convulsions.

Another large group of poisons, the anilines, include many of the more modern drugs, such as acetanilid. Closely allied are different aniline dyes; also phenacetin, antipyrin, etc. In these the characteristic signs of poisoning are somewhat similar to those seen in the phenol group, but in the more pronounced ones of this series the main changes occur in the blood. There is blueness of the skin and lips, difficulty in breathing, sometimes pinkish to purplish urine, rapid and feeble heart action.

Alkaloidal poisons are numerous. The commonest forms of poisoning from these—the most powerful poisons—are morphine (opium, laudanum, paregoric), strychnine (nux vomica), atropine (belladonna), cocaine (coca), aconitine (aconite), and nicotine (tobacco). In acute cases of opium poisoning the classical symptoms are drowsiness, coma, small pin-point pupils, loss of pain, slow breathing (6 to 8 to a minute), moist skin, dry mouth, rousing with more or less active consciousness, and quick relapse. Strychnine poisoning causes twitching of muscles, cramps, irregular muscular movements, convulsions at slightest jar or touch, fixation of muscles of breathing, with cyanosis. Belladonna poisoning shows wide-awake, restless consciousness, sometimes active, busy, delirium; dry mouth, skin hot and flushed, pupils widely dilated and paralysed to light and accommodation; rapid, feeble heart, and rapid respiration.

Another group of poisons—the glycosides—is characterized by a great similarity in action. Many of these are used in medicines, and some were used on arrow-points by wild natives. This group contains digitalin (digitalis), strophanthin (strophanthus), convallarin (lily-of-the-valley), bryonin (bryonia), apocynin (dog-bane), oleandrin (oleander), scillain (squills), etc. They are all heart poisons. They first quicken the heart, then slow and regulate it, hence their usefulness in many heart diseases; but in overdose they paralyse the heart.

Toxic albumins form a group of special character, and all are very violent. Some are of vegetable and others of animal origin. The most important are abrin (in jequirity seeds), ricin (from the seed-coats of the castor-oil bean), phallin (in poisonous mushrooms), rattlesnake poison, cobra poison, and the poison of lizards, etc.

4. Death by Freezing

Let us now examine a few of the numerous cases that have been reported in which individuals have frozen to death—and almost died, but afterwards recovered to tell of their sensations. We have objective evidence in the former case; subjective in the latter; and needless to say, the latter is by far the more valuable. The objective indications of freezing are surely too well known to need re-statement—the whitening and deadening of the parts; the numbness and stupor which gradually creep over the body—all this can be observed by an outsider. But let us turn to the subjective or interior state, meanwhile, and see in what that consists. A few summarized cases will do for our present purposes. Thus, one person who almost froze to death, but subsequently recovered, writes as follows:—

"The process of dying, arising from freezing and the consequent benumbed feelings and sleepy sensations, is undoubtedly painless. When a person feels exceedingly drowsy, he dislikes to be disturbed, and, when freezing, he seems to be oblivious to the great dangers that threaten him. This, as a natural consequence, arises from the weakness of the will—however that may be caused—and a disposition to quietly submit to the domineering actions of the feelings. Sleepiness caused by freezing is enervating; the brain ceases to be stimulated in the proper manner, and the vague dreams, accompanied by strange illusions, succeed the active energies and thoughtfulness of the mind. In extreme cold, the physical system is outside of its sphere or its normal healthy element, the same as it would be if thrust into water, in a well where gas would stifle it, or in an oven, where it would gradually roast. . . . Freezing may be denominated the 'sleep of death,' for a sleep, calm and peaceful, precedes the final dissolution, and the awakening can only be in that region towards which all are tending. Of course such a death, after the first tingling sensations have quietly passed away, must be painless. Few, however, seek that method to commit suicide. The first exposure to the cold is very disagreeable, and those intent on self-murder hesitate before they expose themselves to its initiatory influence—hence they oftener use the pistol, or poison, or jump into the water."

Another, narrating the sensations while "dying," thus describes them:—

"Thousands of coloured lights danced before her eyes;[1] the roar of a thousand cannon was sounding in her ears, and her feet tingled as if a million needle-points were sticking into them as she walked. Then a feeling of drowsiness came over her. A delightful feeling of lassitude ensued—a freedom from all earthly care and woe. Her babe was warm and light as a feather in her arms. The air was redolent with the breath of spring. A delightful melody resounded in her ears. She sank to rest on downy pillows, with the many coloured lights dancing before her in resplendent beauty, and knew nothing more until she was brought to her senses."

Still another writes:—

"The bitter cold does not chill and shake a person, as in damper climates. It stealthily creeps within all defences, and nips at the bone without warning. Riding along with busy thoughts, a quiet, pleasurable drowsiness takes possession of the body and mind, the senses grow indistinct, the thoughts wander, weird fancies come trooping about with fantastic forms, the memory fails, and, in a confused dream of wife and home, the soul steps out into oblivion without a pang or a regret."

There are several distinguishing marks between *rigor mortis* and a body that has been frozen to death. In cadaveric rigidity the skin is soft and pliant; in the frozen body it is not. In cadaveric rigidity, when we move the limbs there is no sound; but in frozen bodies a crackling sound is emitted.

[1] It is related in the third person.

5. Death by Starvation

The length of time it is possible to live without food varies greatly in warm- and cold-blooded animals. Chossat found that in different warm-blooded animals death resulted when the body had lost about 40 per cent. of its normal weight. He found that in animals undergoing starvation the symptoms observed during the first half or two-thirds of the period are those of calmness and quietness; the temperature then becoming elevated, restlessness and agitation prevail; and when life is terminated by the rapid fall of the temperature, stupor supervenes. There can be no doubt that individuals can subsist without food far longer than is usually supposed—many cases of sixty-day fasts, and even fasts of longer duration, being recorded from time to time in various medical works. These cases have been studied from the point of view of starvation pure and simple; and, when the individual is normal at the time of commencing such starving process, there can be no doubt that the effects noted would be such as are indicative of harmful or destructive results to the organism.

Starvation only occurs, as a matter of fact, after a much longer time than is generally supposed. A man may exist for two or even three months without food, under certain conditions; and, during the first part of that time may even receive benefit from the abstinence. That is while *fasting,* however, and not during the period of *starvation.* The two processes are very different, as I have elsewhere tried to show at considerable length. (See *Vitality, Fasting and Nutrition,* p. 564.) When fasting ends, starvation begins, and that is a very different thing. Then the tissues shrink, the body wastes, and the mind becomes impaired. The moral faculties also become blunted, there is good reason to believe; cases of cannibalism among civilized people would seem to indicate this. Dr. N. E. Davies, writing some years ago on this question in the *Popular Science Monthly*, said:—

> "Reasoning by analogy, we find that, in many cases of bodily disease, the state of the mind is the first indicator of the mischief going on in the system. Take even such a simple thing as indigestion, which, as every one must know, is only a manifestation of a deranged stomach, and what do we find? That the lowness of spirits induced by the infection may vary from slight dejection and ill-humour to the most extreme melancholy, sometimes inducing even the disposition to suicide. The sufferer misconceives every act of friendship, and exaggerates slight ailments into heavy grievances. So in starvation, the power of reason seems paralysed, and the intellectual faculty dazed really before the functions of the body suffer, or even the wasting of its tissues becomes extreme. Such being the case, the unfortunate individual is not accountable for his actions, even if they be criminal in character, long before death puts an end to his sufferings."

It should be noted, however, that newer researches and observations have tended greatly to modify these earlier views. It has now been shown that man can not only *live* for a far longer period of time than was formerly supposed—as the cases of MacSwiney and other "hunger strikers" prove conclusively—but also that the mental and moral faculties may remain practically unimpaired throughout the period of starvation, provided the *Will* be sufficiently strong. In their cases, as we know, all of them showed a high standard of moral and mental stability throughout, and although MacSwiney suffered from spells of delirium during the latter part of his period of starvation, his sanity and steadfastness of purpose were unimpaired. Evidently, therefore, death from starvation depends, with regard to many of its symptoms, upon the character and the mentality of the patient. Even in Death, the human Will may be supreme!

6. Death by Asphyxia and Drowning

In asphyxia there is more or less complete loss of consciousness, because of imperfect oxidation of the blood. The symptoms may be developed rapidly or slowly. In sudden occlusion of the air passages, such as is caused by a foreign body in the larynx, or compression of the throat, as in hanging, there is usually a quiet period of from twenty to thirty seconds, after which respiratory movements both of inspiration and of expiration follow. These gradually increase in frequency and depth until, in about a minute, powerful expiratory convulsions occur; convulsive movements of inspiration are also produced, but these are usually milder in character. A period of exhaustion sets in, the respiratory movements become slower and more irregular, and gradually cease. During this period the face has become pallid, and then deeply cyanosed and flushed, the lips blue to purple, and the body temperature, at first increased, gradually diminishes. The blood-pressure is at first increased, and then falls gradually to zero. Unconsciousness develops about a minute after the occlusion, although there is great individual variation; the sphincters relax and the urine and faeces are passed. There is a loss of muscle tone, and the reflexes are abolished. In asphyxia both lack of oxygen and increase of carbonic acid gas in the blood are important factors.

Among the most important phenomena that are to be observed are the following:—The cooling of the body is generally slower in all forms of death from asphyxia. Then in asphyxia the blood is always very fluid, and few clots are found in the heart or great vessels. Owing to this fluidity, hypostasis is well marked. The blood is generally very dark in colour. The next point is the congested condition of the lungs. Small patches appear at the root of the lungs. Tardieu considers them distinctive of suffocation, but in this he is probably too dogmatic.

In *strangulation* we have the circulation to and from the brain impeded. The face is commonly pale and placid; prominent eyes are not uncommon. Protrusion of the tongue appears frequently; the hands are often clenched.

"Death by asphyxia begins at the lungs, almost simultaneously paralysing the muscles of the body. The victim is deprived of the power of action, while still retaining consciousness. Not even an outcry is possible, and death approaches inch by inch—relentlessly entangling the agonized victim in its skeins, from which there is no escape, unless timely help arrives before the last stage in the passive struggle. While still conscious, the brain, in its attempts to break the chains of death, pictures the past and present in vivid colours, flashing like lightning over the memory, which still has a conception that the end is coming."

This picture-forming faculty of the mind at the moment of death is supposed to be most common in cases of *drowning*. The past will come up before the mind with marvellous rapidity and detail, at such times; and the latter would seem to know no limitations of time or space. This is a most significant fact. In cases of death from strangulation, asphyxia, etc., the blood becomes nearly *black*, by reason of its passing through the lungs several times without aeration. When death results from the taking of opium, and certain other drugs, it is said that consciousness of the entire body is lost before the senses or intellect become dulled; but this seems to me very doubtful.

7. Death from Shock

It is asserted that, in many cases of this character, the patient may be brought back to life by careful and persistent treatment—on the line of "first aid to the injured." Resuscitation may be affected, it is claimed, just as in cases of drowning, in many instances. Shock of this character may produce "death" in either one of three ways: First, by producing destructive tissue changes, when death is absolute; second, by producing sudden arrest of the respiratory and heart muscles through excitement of the nerve centres, when death is only apparent—in other words, animation is merely suspended; or, third, by a temporary exhaustion of nerve force—the result of a violent, sudden, and excessive expenditure of it. The subject may be aroused from this syncope if efforts at resuscitation are not too long delayed. In cases of this character, the oxygen treatment is sometimes very efficacious. Electricity or even cold water may be applied with great success in cases of "shock."

The symptoms of shock vary greatly according to the type of cause and the individuality of the patient. Sometimes the symptoms begin at once; under other circumstances the alleged results may be delayed for a long period. Surgical shock is, perhaps, one of the most severe. The symptoms of all forms of shock are very similar. The face usually becomes blanched and pale, the body becomes cold, and is covered with a clammy perspiration; the hands and feet usually become icy, the brain seems to be in a whirl, consciousness is lost, or much clouded. The pulse is usually quickened; the eyes sunken and listless.

All such cases bring before us very forcibly the possibility of body resuscitation. Various devices have been employed to this end, some of which have been mentioned above; and there are yet others—artificial bellows, methods for restoring the circulation, and similar means—besides the well-known methods classed under "first aid to the injured." Injections of certain saline solutions into the veins have sometimes been accompanied with remarkable results. Perhaps the most powerful of all these measures, however, is *cardiac massage*. It has been asserted that, by this means, a heart has been made to beat after having stopped for several minutes. A long series of experiments should be conducted along these lines, and the results made public. So far as I know, no experiments have ever been made in which the efficacy of *suggestion*—hypnotic or other—has been tried, at the moment of death.

(I must except Poe's tale, *The Case of M. Valdemar*—a work of pure fiction, as Poe afterwards admitted.)

8. Death by Electricity and Lightning

While it might almost be said that the body died first in cases of freezing, and that consciousness was only extinguished slowly at the end, precisely the reverse of this is present in all cases of electrocution, or death by electricity. In such cases, the consciousness is certainly obliterated at once, but the cell-life of the body as certainly persists for a long time after the electrocution takes place and the body ultimately putrefies, as in other cases. But these are questions which require much investigation in order to settle them satisfactorily.

It has been asserted that a large proportion of cases of electrocution might be resuscitated if the proper measures were adopted at once. This may be very true in certain instances, but it is certainly not true in the great majority of cases, as electrocution is performed today. A most remarkable instance was reported, however, from Pittsfield, Mass., where, on 23rd October 1894, James E. Cutter, working in the testing-room of the Stanley Electrical Manufacturing Company, accidentally received 4600 volts of electricity, and was afterwards resuscitated by two fellow electricians, who treated him in the same manner as one would be treated who had become unconscious through drowning. At the end of seven minutes he recovered. Writing of the incident, he afterwards said:—

"For a brief instant there was a sensation as if I were being drawn downwards by the arms, and then everything became blank. For several minutes there was no sign of life. . . . Then slowly I began to regain consciousness and to make incoherent remarks about the accident. Half-an-hour afterward I could recall every incident before and after the seven minutes' interval, which was a total and painless blank. The accident occurred about ten o'clock in the morning. For the remainder of the day I was quiet, but on the following day I was around as usual. I have experienced no ill effects other than the scars from the burns, one of which went to the bone."

As is well known, one of the most important safeguards of the human body against the passage of electrical currents through it is its high degree of resistance. This degree of resistance, however, is subject to a considerable amount of variation. If the skin is dry, the resistance is from five to twenty times as great as when the skin is wet. From what is known of the amount of electrical current necessary to cause death in man, it is probable that 1600 volts of electro-motive force of a continuous current is sufficient to bring about this end, and that an alternating current of half this voltage would probably be fatal. In fact, the general deduction has been drawn from the experiments conducted in electrocution work at the Sing Sing prison, that no human body can withstand an alternating current of 1500 volts, and 300 have produced death, while for the continuous current it may be necessary to pass 3000 volts, in order to bring about fatal results.

The number of deaths from lightning is larger than would be ordinarily supposed. The injuries produced by it often simulate external violence. The clothes are frequently torn off the body, and part of the clothes or the bodies themselves thrown great distances. Again, we may find metallic things about the body fused, and any iron thing is rendered magnetic. Marks like prints of trees or foliage may occasionally be found on the body after it has been struck, as though photographed upon it. This is an undoubted fact. Many of these caprices of lightning are very striking. At one time a stroke of lightning set fire to a man, and he blazed like a sheaf of straw; at another it reduced a pair of hands to ashes, leaving the gloves intact; it fused the links of an iron chain as the fire of a forge would do; and, on the other hand, it has killed a huntsman without discharging the gun which he held in his hand; it has melted an earring without burning the skin; it has consumed a person's clothing without doing him the slightest injury, or perhaps destroyed his shoes or his hat; it has gilded the pieces of silver in a pocket-book by electro-plating from one compartment to another without the owner being aware of it; it has demolished a wall six or eight feet thick in a moment, or burned a château a hundred years old, yet it has struck a powder factory without causing an explosion.

Dr. John Knott, writing in the *New York Medical Journal,* says:—

"The materialistic nineteenth century does not fail to find an explanation in what has since been recognized as *return shock.* Every substance capable of conducting the mysterious electrical fluid, on being placed in the vicinity of an electrified ('charged') body—and not connected with the same by a conducting medium—becomes charged with electricity of the opposite kind, and to approximately the same potential or electromotive force. In accordance with the physical necessity which determines this process, a man may stand within a moderate distance of a thunder-cloud, which holds an enormous charge of, let us say, positive electricity. In such position, his body necessarily becomes charged with negative electricity, by the influence of what is known as *induction.* While the state of equilibrium is maintained, without any abrupt disturbance, he feels no ill effect or inconvenience whatever. But when that cloud discharges its electricity in an opposite direction, the inductive influence instantaneously ceases; the induced negative charge is (in the same instant) neutralized by drawing an equal quantity of positive from the 'universal reservoir' of the earth. The shock corresponds in intensity to that producible by the discharge of the cloud itself, and passes through the nervous system with such effect that the individual drops dead instantaneously, and without a single trace of injury on or around his person."

In cases of direct contact with the lightning flash, burns, more or less extensive and penetrating, have been noticeable; but as a rule there is nothing very remarkable about them. One of the most characteristic signs of the *post-mortem* conditions in cases of death by lightning is that when the shock has been direct and very powerful, the blood fails to coagulate after the normal fashion. (After electrocution, imperfect coagulation of blood has been noticed, giving rise to the supposition that the subject is not really dead. Such, however, does not follow, as we have seen.)

9. Death by Spontaneous Combustion

Dr. Trall, in his *Hydropathic Encyclopedia,* vol. ii, pp. 179–80, says:—

"This is a condition of general combustibility of the body, produced by the use of alcoholic drinks. Examples of spontaneous combustion, as having occurred in persons long accustomed to the immoderate employment of spirituous liquors, are too well authenticated to be longer doubted. The condition of the body liable to this strange phenomenon may properly be called *alcoholic diathesis.* In a majority of the cases recorded, females advanced in life are the subjects of the malady. In some cases the self-consuming flame has arisen without any obvious exciting cause; but in others a fire, a lighted candle, the heat of a stove, or an electric spark, has ignited the inebriate body. It is a remarkable fact that the flame which decomposes and reduces every fragment of the bodily structure to ashes does not essentially injure the common furniture or bedding with which it comes in contact; and more marvellous still is the statement that water, instead of quenching the fire, seems rather to quicken it!"

Again, Dr. Joel Shew, in his *Family Physician*, pp. 717–18, says of this condition:—

"That the living body becomes at times, in consequence of long-continued intemperance in the use of alcoholic drinks, liable to combustion, easily excited and spontaneous, is abundantly proved. The condition, however, is a rare one. Some doubt the facts, but, as a French writer has observed, 'it is not more surprising to meet with such incineration than a discharge of saccharine urine or an appearance of the bones softened to a state of jelly.'

"This condition of the system will appear more remarkable when it is remembered that in all other states, whether of health or disease, the body is with difficulty consumed by fire, even at a high temperature. . . .

"This phenomenon seems to have taken place for the most part in the night time, and when the sufferer has been alone. It has usually been discovered either by the fetid, penetrating scent of sooty films, which, as we are told, have spread to a considerable distance; or by the blue flame that hovers over the body; or the unnatural heat, which, however, is not very great. The patient in all cases has likewise been found either dead or so far consumed that life appeared to be extinct; and in no instance has recovery been known to take place after the appearance of this most singular of all pathological states."

There is practically no belief in spontantous combustion in these days, but it is admitted that in certain cases the body may acquire preternatural combustibility. This is founded on the assumed fact that much of the body has been found consumed while surrounding objects are not much consumed. In nearly all well-authenticated cases there has been some source of fire near, probably setting the clothes on fire, usually when the sufferer was habitually drunk and so could not help himself.

It is of interest to note, in this connection, that a case in which somewhat similar phenomena occurred after death recently came under my own observation. The patient was a child who had died of acute indigestion caused by eating a large quantity of chestnuts without properly masticating them. After a day spent in the chestnut grove the child returned home, and about three o'clock in the morning died in terrible convulsions. As this occurred in the country, the neighbours volunteered to prepare the body for burial, and it was while the work of making the shroud was in progress that it was discovered that the entire body was, to all appearance, on fire. The glow extended from the head to the feet, and could not be extinguished, although it finally died out, disappearing altogether. While the heat from the bluish flame which enveloped the body was quite perceptible, it was not sufficient to burn the body or even set the bed on fire; and yet, when the corpse was removed from the sheet on which it had been placed, it was found that the latter was scorched in such a manner that the outlines of the human figure could be plainly distinguished. In this case, it will be noted, alcohol played no part in the production of the phenomenon, but there can be no doubt that the chemical changes were similar in character to the cases previously cited.

In another case known to me, a kitten remained upon a rug for an hour or so after death; when removed, a distinct outline of the kitten's body was left upon the rug. This was doubtless due to some chemical action, but as to its precise nature I do not feel competent to speak.

CHAPTER IV

THE SIGNS OF DEATH

MANY years ago the Marquis d'Ourches offered, through the Paris *Académie de Médicine,* two prizes, one of twenty thousand francs, the other of five thousand francs, for some simple, certain sign of death. The secretary, Dr. Roger, reported on the competition. One hundred and two essays were sent in, but none was deemed worthy the first prize; The second was divided between six competitors. Five hundred francs was given to M. de Cordue for his observations on the effects of the flame of a candle on the pulp of the finger. M. Larcher was rewarded for his observations on the eye after death. (As the result of examining nine hundred patients, he found the occurrence of a shaded or greyish spot, first on the outer portion of the sclerotica, and gradually involving the whole surface.) M. Poncet received an honourable mention for his observations on the discolouration of the fundus of the eye; M. Molland, for his observations on cadaveric lividity; and MM. Bouchut and Linas for their observations on the temperature of the body. But nothing definite or decisive was discovered.

Passing in review the various signs of death, M. Brouardel has this to say:—

"The *combination* of signs of death gives us almost complete certainty of death. . . . But I believe that it is right to remain in a state of philosophic doubt; we know that apparent death may last for a longer or shorter time, and that in three cases at least . . . persons considered to be dead have been called to life. . . . The verification of death should therefore always be entrusted to a physician, who alone is competent to estimate the value of the different signs that we have just been examining. . . . I believe that accidents will then be, if not impossible, at any rate infinitely rare, and I am obliged to add that though there is a great improbability of a living person being buried alive under those conditions, in which actual death is, or rather is not, complete,—still, it is impossible to assert that the direful contingency might not happen" (pp. 61, 62).

1. General Signs

Let us, then, see what these signs are, which are supposed to render death certain, and thus prevent these unfortunate "accidents," or this "direful contingency."

In death, *intelligence* is absent; but so it is in trance and syncope.

In death, *insensibility* is complete; but it is also practically complete in certain cases of hysteria in which there is complete anaesthesia. Surface insensibility is complete, and the patient does not react to the most painful tests, on occasion. All sense of *hearing* and *smell* are also absent. The *eye* presents some very interesting tests. It was noticed that there was an immediate lessening of the tension of the globe of the eye, just after death, owing to the fact that the blood-vessels were emptied of blood. But this proves, merely, that the heart has stopped beating—not that death has taken place; and we know that persons can often be revived long after the heart has ceased to beat. Bouchut contends that atropine and eserine have no effect after death. The pupil dilates at the moment of death, but afterwards returns to its normal condition and size, and the iris is thrown into folds. It is also asserted that the eyeball is harder after death than during life.

One very characteristic sign is the *sclerotic speck* that appears after death; the conjunctiva also assumes a brown hue. Commenting on these signs, Dr. Hartmann wisely remarked, "All these signs prove that the circulation has stopped; not that it cannot be started again."

It will be of interest to refer here to a peculiar fact, the explanation of which is still somewhat uncertain, which caused a tremendous sensation some years ago when it was first made public. It was announced at the time that in persons dying suddenly the eye preserved the impression of whatever object was in front of it at that moment. It was suggested that murderers might be traced in this manner—since it is to be supposed that the murderer would be the last object seen by the murdered man, in most instances. The case was somewhat overstated, and many persons totally disbelieve in the possibility of the fact at all. There is, however, some ground for the belief. Kühne of Heidelberg placed a grating in front of a rabbit, then killed the animal rapidly, removed its eye, exposed the retina, and photographed it. The cross-bars of the grating were clearly seen in the print. In the case of a more complicated object, such as a table or a chair, the outline was much more blurred and indistinct, but yet recognizable. In such cases the animal must be killed immediately, and the retina photographed very soon after death. For these reasons, it would be difficult to obtain definite results in a human being. Certainly, no visible trace of any scene would be found on the retina twenty-four hours after the death of the subject. This is a question of great importance that should be followed up closely; but, until some of the prejudices of the public are overcome, it is unlikely that any definite results will be obtained in this possibly fruitful field.

At death the *immobility* of the body becomes pronounced, and the lower jaw falls on to the breast. But these signs are not constant, and it has been pointed out that in tetanus and in hysteria the mouth may remain closed. Complete *rigidity* of the corpse may sometimes be found before *rigor mortis* supervenes. After death, as the body cools, the muscles, especially of the face, continue to contract in odd ways, and sometimes the face will be pulled into various shapes, and give the appearance, perhaps, of the patient having died in the greatest agony. Such may not have been the case at all; the death may have been perfectly painless. Richardson attached considerable weight to the fact that live bodies usually respond to an *electric stimulus,* while dead bodies do not. But this test also has been found inconclusive.

Respiration ceases at death; yet the respiratory test is quite variable in its results. In some cases the patient may be in a trance, and appear not to breathe at all, and yet be alive. On the other hand, a patient may be dead, and the gases moving about within his body give the appearance of life. The old test of holding a mirror to the lips is known to all; the idea of placing a glass full of water on the epigastrium of the patient is not so well known. If this overflows the patient is supposed to be alive; if not, he is dead. The test is inconclusive for the reasons indicated above.

Brouardel, in his excellent manual on *Death and Sud-*

den Death, thus enumerates the sources of error in attempting to assure oneself of the fact of death by observations upon *the circulation:—*

"Bouchut, who has studied all these questions with great care, has rightly said that one must not be satisfied with feeling the pulse, but must go higher and consult the heart also. In a memoir published by him, and submitted to the Academy of Science, he states that an interruption of the action of the heart, lasting for two minutes, was sufficient to render the diagnosis of death certain. Andral, who was appointed to report on Bouchut's memoir, believed that this interruption should be prolonged for five minutes. Later on, he was obliged to acknowledge that even this length of time was inadequate, since in the interval he had met with a woman who returned to life some hours after the action of the heart had ceased to be perceptible; it is true that a few contractures could be perceived from time to time, but they vanished to reappear later.

"Bouchut thinks that the heart should be listened to for half-an-hour. There are at least two sources of error here. You cannot listen to a heart for half-an-hour continuously. Try to do so; in five or six minutes you will hear buzzings and murmurs of all sorts, and at last you will hear the beating of *your own* heart. A second source of error is as follows: When an animal is dying, and you practise auscultation, you hear very plainly the two sounds of the heart, then only one sound, which presently disappears also. If the animal is opened the heart is found still beating. Therefore, it is essential that the heart should beat with a certain degree of energy in order that its beats should be heard" (pp. 50, 51).

He also points out that the acuteness of hearing is not alike in all.

If the absence of the heart-beat cannot be considered a certain sign of death, perhaps some of the other signs connected with the circulation might. If the *vein* be opened immediately after death no blood will issue therefrom; but blood will issue in the course of a few hours if the wound be left open. The arteries contract, and force the blood through the capillaries into the veins. Further, the gases formed within the body force the blood to the surface, so that, if the skin be cut, blood will sometimes flow. This was the origin of many of the stories of "vampires" to which I refer elsewhere.[1] *Coagulation* is also a very uncertain sign. *Ligature of the finger, cupping* and *leeching,* have been resorted to; but the same objection may be raised to all, viz., the fact that the heart's action has ceased does not guarantee that it cannot be set in motion again.

After death, little livid spots appear on the surface of the body. They are known as *cadaveric sigillations* or *lividity,* and are caused by the exudation of blood into cellular tissue from the veins. It is an almost invariable sign. Dr. Molland, who examined 15,146 cases, never found it absent once. Nevertheless, it may be absent in cases where there has been abundant hemorrhage before death; and, on the other hand, they may appear before death in certain cases—in cholera, uraemia, and asphyxia. This sign is also, therefore, inconclusive.

The *temperature post-mortem* has been considered a very important sign; but it is a very uncertain one. When the surrounding temperature is high, the body may take a very long time to cool, though death may have taken place; and certain diseases also hinder the cooling of the body. On the other hand the body may cool considerably in trance, and certain states of a kin-

1 See Appendix A.

dred nature, and yet life may be preserved and revived.

In slow deaths cooling is a gradual process, and varies much in rapidity. The trunk may remain warm, while the limbs are cold. The cooling is slow if the body is covered with warm clothing, or bed clothes. Wool is a bad conductor. The bodies of young persons, which have generally a subcutaneous layer of fat, take longer to cool than those of thin, old persons. In wasting diseases the heat is low in the last hours before death. It has been supposed that cooling takes place more rapidly in cases of death from hemorrhage, but this is rarely true. In all cases of death by suffocation, cooling seems to be retarded. Casper's rule as to the cooling of the body is as follows: "A body found on the highway with clothes on (the air being at a medium temperature), still warm, has been dead probably not more than three hours. A body found in bed and still warm has been dead at most for ten or twelve hours."

Another sign of death that can sometimes be obtained is the following. A patch of skin is removed, and, in the course of some hours, the exposed surface will become *parchment-like* in appearance, and will yield a sharp sound when tapped. I do not know if this has ever happened in a case of trance; and we have, consequently, nothing to guide us in this respect.

A sign that was for long considered certain was that of *burning* or *blistering the body.* If a live body be burned, a blister will be raised, surrounded by a reddish areola. In dead bodies this is supposed not to exist. But is that the case? M. Brouardel states that blisters may very readily be raised on dead bodies:—"Let a drop of melted sealing-wax fall on to a limb that has just been amputated, and you will succeed in producing a blister.'' The test of burning is therefore a doubtful sign.[1]

Dr. Franz Hartmann, in his excellent manual, *Buried Alive,* has summarized quite exhaustively the various signs of death. I abridge his account of those tests that other authors have omitted to mention.

Immobility of a needle stuck in the pericardium:—This indicates that the heart has ceased to beat; not that the person is beyond recovery.

Emptiness of the central artery of the retina; disappearance of the papilla of the optic nerve; discoloration of the choroid and retina; interruption of the circulation of the veins in the retina; emptiness of the capillary vessels:—All these signs are open to the objection just pointed out.

Corpse-like face; discoloration of the skin; loss of transparency of the hands:—''These signs are now so well known to be delusive, as to require no further attention.'' *Emptiness of the temporal artery:*—This only indicates that the heart has lost the power to send the blood to that artery; but it is no sign that it may not recover its strength. *White and livid colouring at the points of the fingers:*—An antiquated and misleading sign.

Relaxation of the sphincters and the pupil; glazed eyes and haziness of the cornea; insensibility of the eye in regard to the action of a strong light; bending of the thumb towards the palm of the hand:—All given up nowadays as unreliable.

[1] The author just cited states that he has found an excellent way of reviving those in syncope; it is to place a hammer just dipped in *very hot* water on the epigastrium. Patients nearly always revive. It is doubtful if this would succeed in every case, however—especially where the vitality is very low; and indeed the author intimates that it would not.

Disappearance of the elasticity of the muscles—also takes place in dropsy and other diseases.

Non-coagulability of the blood:—Unreliable; in scurvy and certain other diseases the blood remains incoagulable for several days.

Absence of a humming noise in the auscultation of the finger points:—Unreliable. If the finger is not held in just the right position, nothing will be heard, even if the patient be alive. Further, humming noises, internal noises in the body of the physician, etc., are apt to be mistaken for the sounds going on in the body of the patient.

Galvanism has been considered insufficient to furnish a test that is certain. *Irritability* is extinguished first in the left ventricle; then in the intestines and stomach, next in the bladder, afterwards in the right ventricle, then in the oesophagus, and after that in the iris. The muscles of the trunk finally give way—the extremities and the auricles. The collapsed edge of a wound in a dead body, in distinction from a gushing wound in a living one, is the result of a peculiar irritability—the extinction of which is one of the indications of death. Flaccidity is an uncertain sign of death; putrefaction is unequivocal.

Within recent years, several novel tests have been devised. *X-ray* machines have been employed to ascertain whether any vital action was taking place within the body. It was found that, if all the internal functioning had come to a complete standstill—bowels, liver, lungs, heart, etc.—the shadow cast on the screen would come out clear and distinct; if, on the other hand, some of these organs were working (and consequently *moving*) the outline or shadow would be blurred and indistinct.

I do not know to what extent this test has been carried; and its value and reliability would depend (1) upon the clearness of the shadow; and (2), upon the extent to which the internal organs can suspend their functioning, in such states as trance, and yet life be present, or possibly recalled. We must always remember that the entire vital machinery might stop, for a short time, and yet be enabled to resume its functioning. This fact must be taken into consideration when discussing this test.

On the other hand, Dr. Elmer Gates has published an article in the *Annals of Psychical Science* (June, 1906), entitled, "On the Transparency of the Animal Body to Electric and Light Waves: As a Test of Death and a New Mode of Diagnosis, and a Probable New Method of Psychic Research." He says in part:—

"Several years ago . . . I discovered that certain wave lengths of *electric waves* (not X-rays or ultra-violet light) pass more freely through the body of a dead than a living organism, and I proposed this as a test of death. This greater transparency at death I found to be due to the absence of the normal electric currents, which are always present in functionally active nerves and muscles. . . . When the body is alive, it is a bundle of electric currents, and electric waves cannot pass through these currents; but when they cease, at death, the body becomes transparent to electric waves."

How far these electric currents would be reduced in trance and kindred states, is a matter for further inquiry. The objections previously raised must not be lost sight of in this connection.

There is yet another test of death of a somewhat "occult" character, which its votaries declare infallible! It is the following:—

"*The Aura after Death.*—It will readily be understood that death produces an immediate great change in the human auras. All the higher principles, together with the auric egg that envelopes them, disappear, leaving the doomed material body with only its lifelong and inseparable etheric double floating over it; the caloric aura gradually ceases with the disappearance of animal heat; the pranic aura, which had begun to fade before the actual dissolution, turns to an ashen-grey light; all the electric emanations, already broken up during the sickness, cease; the magnetic flow alone continues, though in a sluggish and stationary manner; the Tatwic ribbons lose their colour, leaving only dead, colourless lines, as in mineral matter, whereby it can be said that the auric manifestation which remains around the body is only that which belongs to the dead material compounds, until decomposition sets in. Then the auric effluvium again becomes alive, and assumes the aspects and hues of the new lives that issue out of death. Thus, the study of the human aura will bring out new and more reliable signs of real death, because to a psychic sight, the aura of a person in coma or cataleptic trance—however well this may otherwise simulate death—will never be mistaken for that of a body in which life is really and positively extinct. . . . "[1]

Without discussing the reality of these phenomena in this place, it may only be said that the difficulty of finding a seer possessing the requisite psychic sight might be sufficiently difficult to render this method of diagnosis impractical under all ordinary circumstances! Of course, such theories would have to be rigorously demonstrated before science could even tolerate them, in a life and death problem such as this.

[1] *The Human Aura*, by A. Marques, pp. 55, 56.

2. Odor Mortis; or, the Smell of Death

In the Cincinnati *Clinic* of September 4, 1875, was published a paper on "Odor Mortis; or, the Smell of Death," read by Dr. A. B. Isham before the Cincinnati Academy of Medicine, August 30, 1875. The paper was based upon observations made while he was in one of the surgical wards of the Stanton Hospital, Washington, during the summer of 1863, as well as upon instances in which the "odor" had been met with in private practice. The character of the odour was muskiferous, yet it appreciably, though almost indescribably, differed from that of musk. In this paper he presented two recent instances where this odour attracted notice, together with some new observations concerning it.

Instance 1.—July 13, 1878, on the eve of Dr. Bartholomew's departure for Europe, Dr. Isham was requested to assume charge of his patient, Mr. ——. The patient was unconscious, with irregular, noisy respiration, with only a feeble trace of pulse, indistinguishable at times, and was dying slowly from effusion within the membrane of the brain, the result of chronic alcoholism. He was with him through the middle of the night, and during this time he noticed upon his right hand a smell resembling that of musk. This hand was exclusively used in examining the patient's pulse and in noting the temperature of the body. Earlier in the night there had been no smell upon it. The left hand acquired the same smell from handling the body, and it was also communicated to the handle of a fan held in the hand. A gentleman from Chicago, who had volunteered as a night watcher, and whose attention had been called to the odour without any suggestion as to its character, promptly distinguished it. The ladies of the household did not use musk, and no perfumery had been in the room or about the patient. Neither had Dr. Isham handled nor come in contact with anything other than the patient from which the odour could have been derived. Death occurred thirty-three hours later.

Instance 2.—About midnight, May 21, 1879, Dr. Isham was called to see Mrs. G. She had several months previously been under his care with acute duodenitis, but with impaired digestion and defective assimilation; but she had subsequently passed into the hands of an irregular practitioner. He found her *in articulo mortis,* with general anasarca, the result of blood dilution. Upon entering the room there was a plainly perceptible musky odour. There was no musk about the house, nor had any other perfumery been employed. Death ensued in about half-an-hour.

The smell, as stated, was closely allied to that of musk, yet the impression on the olfactory organs was more delicately subtle. Besides, there was an indescribable feature pertaining to it which seemed to impress the respiratory sense and trouble respiration—a vague sensation of an irrespirable or noxious gas. To the convalescent loungers of sharp olfactory sense about the wards of Stanton Hospital the smell was familiar, and was termed the *death smell.* It was not uncommon to hear the expression, "Some one is dying, for I smell him!"

It was rare to find the odour widely diffused, and where it appeared to be it was probably due to a continuance of the first impression upon the olfactory organs. As commonly encountered, it has suggested the idea of gaseous aggregation or body containing odoriferous particles possessing an attraction for each other, and so held together. In the hospital ward, while present in one place, it was not experienced in another slightly removed. It also quickly disappeared from the first place—probably moved along by atmospheric waves. The vapour in which the odorous molecules were suspended appeared, in some instances at least, heavier than the atmospheric air. Thus, Dr. Isham had sometimes recognized the smell in lower hallways—the patient occupying the upper portion of the house; and in "Instance 1," already detailed, it was only detected on handling the body. This affords one explanation why it may not claim more recognition. From its heaviness it subsides, and does not enter the nose. Other reasons why it may escape attention are, that the olfactory sensibilities may be blunted by long continuance in an ill-ventilated, bad-smelling sick-room; or the air currents may carry the odour in a direction not favourable to observation.

The only mention of an odour which might be analogous is reported by Dr. Badgely, of Montreal, in a report on "Irish Emigrant Fever." It is thus quoted by Drake in his work on the "Principal Diseases of the Interior Valley of America," as taken from the *British Medical Journal:*—

"I hazard the idea that the ammoniacal odour emanating from the living body, so strong on opening the large cavities and so striking on receiving some of the blood of the vessels —arteries as well as veins—into the hand, were all due to the same condition of this fluid—the actual presence of ammoniacal salts, one of the surest proofs of the putrescent condition of the vital fluid; in fact, to speak paradoxically, of *the existence of death during life.*"

Here the source of the smell is indicated as coming from the development of ammonia in decomposing blood. It is known that musk contains ammonia largely, together with a volatile oil. Robiquet holds that its odour depends upon the decomposition of the ammonia, liberating the volatile matters of the oil. The blood also contains a volatile oil, and it is well known that it possesses odour. This odour may be devoloped by adding sulphuric acid to blood and boiling it. This process was formerly resorted to in order to distinguish blood in questionable cases, but it has been rendered obsolete since the discovery of the blood corpuscles by the microscope. Such a method would be well suited to drive off the ammonia, free from decomposition, together with the volatile oil—to which substance the odour is very likely due.

Originally, Dr. Isham was inclined to limit the occurrence of the manifestation to within a very short time of death. That it cannot be so restricted is evidenced by "Instance 1," when it was noticed thirty-three hours before death. The conditions here were not unfavourable for its development. From the state of circulation, chemical changes were evidently proceeding in the blood, elevating its temperature and liberating those matters to which we would ascribe the origin of the death smell.

Richardson and Dinnis have shown by experiments that ammonia salts added to blood preserve its fluidity by preventing the decomposition of fibrin. This is not without a bearing upon the origin of the *odor mortis.* In gradual death coagulation commences first in the capillaries, and proceeds towards the heart. The escape of ammonia from the blood in the peripheral vessels, liberating the volatile principles and engendering smell, permits local decomposition of fibrin long before the heart has ceased its action.

But Lange has more recently investigated the action of ammonia in living and dead blood. He found that carbonate of ammonia added to living blood was only given off at a temperature of 176° F. to 194° F. When, however, ammonia was added to blood from a *dead* animal, it was evolved at a temperature of from 104° to 113° F. It is well ascertained that in many diseases, just previous to death, the blood temperature is raised above the lowest figure given by Lange. In some diseases, too, the blood falls below the normal bodily temperature. This affords another and principal explanation why the *odor mortis* may not be appreciable. These experiments of Lange also show why this smell is not developed by diseases characterized by great elevation of temperature—simply because the blood has lost none of its vital properties.

Sir William Ramsay has stated that:—

"Perspiration consists of caproate of glyceryl, mixed with free acid, I believe. It does not smell nice; but pure caproates are very fragrant if the right alcoholic base is combined. I fancy that woodruffe and verbena are of the same nature as turpentine, and have probably the same percentage composition. However, so far as I know, they have never been investigated."

3. Rigor Mortis

Next to putrefaction, *rigor mortis* may be considered the surest sign of death that we know. Unless the burial clothes are put on the corpse soon after death, it is almost impossible to get them on at all owing to the stiffening of the body. Yet it is contended by certain authorities that frequently there is no *rigor mortis* whatever. Bichat found that in cases in which an individual had been struck dead by lightning, or had been suffocated by charcoal, there was no *rigor mortis*. When complete, *rigor mortis* is very severe; the body becomes as stiff as a board, and it is next to impossible to bend or flex the arms and legs.

Generally, it may be said that rigor mortis appears in from three to six hours after death. Quite frequently it appears before the bodily heat has passed away. Niederkorn gives us the following table, the result of 103 cases observed by him:—

Rigor mortis within 2 hours after death,	2	cases
From 2 to 4 " " "	45	"
" 4 to 6 " " "	24	"
" 6 to 8 " " "	18	"
" 8 to 10 " " "	11	"
" 10 to 13 " " "	3	"
Total	103	

It is evident, therefore, that the length of time that elapses between death and *rigor mortis* varies considerably. It is asserted that "after poisoning by a large dose of strychnine, *rigor mortis* follows immediately upon the phenomena of contracture with existed at the time the patient died."

"With regard to the *duration* of rigidity," says Dr. Brouardel,[1] "we are also obliged to make allowance for different influences. It lasts on an average twenty-four to forty-eight hours. It may, however, last for a few hours only; at other times, it persists for five, six, or seven days. Our *data* with

1 *Death and Sudden Death*, p. 66.

reference to this subject are very scanty. We know that in exhausted individuals, such as those dying with cancer or phthisis, *rigor mortis* appears early, but does not last long; on the contrary, in an individual dying while in good health, it appears late, and is of long duration. . . . Cadaveric rigidity appears first in the muscles of the lower jaw, then in those of the neck and eyelids, then the lower limbs, and lastly the upper limbs. . . . The muscles of the intestinal walls may present a certain degree of rigidity."

The heart becomes rigid after death also; a fact observed by the illustrious Harvey, and noted by him in his *Second Disquisition.*[1]

When persons die from the result of sun-stroke or heat-stroke, they are already half rigid, and it is stated that the heart becomes rigid immediately upon the death of the body. Vallain states that when he was in Algeria, he opened the bodies of dogs dying from sun-stroke, and, when he cut into the heart, it yielded a sound like that of wood! Generally speaking, *rigor mortis* appears much sooner in a warm and moist atmosphere. Indeed, it has been asserted that it takes just as many *hours* to effect the same result in the summer time as it does *days* in the winter. When the body is fatigued, *rigor mortis* appears much more rapidly.

Dr. Brown-Séquard, writing on this subject, said:

"In rabbits, guinea-pigs, cats, and birds, as well as in dogs, I have ascertained that when they are killed by poisons causing convulsions, the more violent and the more frequent the convulsions are, the sooner cadaveric rigidity sets in, and the less is the time it lasts; the sooner also does putrefaction appear, and the quicker is its progress."[2]

1 Harvey's treatise on the circulation of the blood should be read by every one, as it is a model of sound, logical argument.

2 Quoted by Savory, *Life and Death*, pp. 190, 191.

What is *rigor mortis?* What is its nature? In what does it consist? This has been a very vexed question; and only of late years has it been satisfactorily settled. Kühne believed that it was due to the coagulation of myosin, an albuminous substance contained in the muscular tissue. Brown-Séquard objected to this, that no amount of such coagulation would account for the facts. Microscopic examination of muscles has frequently revealed no structural difference whatever between those in a state of rigidity, and those that were flaccid. Some observers ascertained that an acid reaction was found in the muscles at such times; and concluded that rigidity was due to the conversion of alkaline substances into acids; but Achtakaweski has proved that in tetanus the muscles are not rigid, and that the injection of an alkali into the muscular tissue does not prevent rigidity. It has even been ascertained that rigidity will take place as usual, even if all posthumous circulation be cut off! Brown-Séquard removed the spinal cord from an animal and found that no rigidity resulted. His researches, however, have been largely disproved by recent experimenters.

While much still remains uncertain, it is now generally admitted that *rigor mortis* is the first stage of putrefaction—of which we shall presently treat—and is hence the result of bacterial decomposition. Herzen proved that there is found in the muscular tissue of a dead animal, an acid, which he called "sarcolactic acid." By injecting some drops of this acid into the muscles of dead animals, he caused *rigor mortis* to appear in cases which had not as yet exhibited it. *Rigor mortis* is doubtless the result of certain micro-organisms, which secrete toxins in the muscular tissue, causing *rigor mortis* in this manner. The subject will become more clear when we consider the phenomena of putrefaction. To this we accordingly turn.

4. Putrefaction

The phenomena of putrefaction are of great interest and importance, since they frequently enable the practitioner to tell almost exactly how long a certain body has been dead, and for that reason are of great value to forensic medicine. The subject may appear an unpleasant one to many readers; but, rightly considered, it is not so, and affords a field for very interesting experiments and important deductions. Bear in mind the fact, that putrefaction is merely the process of returning the body to the native, mineral elements, and there should be no objection to studying this process from the scientific point of view. Remove from the mind the idea of a "corpse," and replace it by the following: here is an organic compound; let us watch its gradual disintegration and return to mother earth!

It has been proved that if a body be perfectly preserved from the air, or oxygen, it will not, *caeteris paribus,* decay or putrefy at all. Pasteur experimented with blood and urine, among the most fermentable and putrescible of all organic fluids. These fluids he sealed up hermetically in glass tubes. Although these tubes are in his laboratory yet, having been placed there in 1854, there is today not the slightest trace of putrefaction in any of them. The presence of *air* is therefore necessary, in order that putrefaction may proceed. Why is this?

When a body dies, three different and distinct sets of micro-organisms occupy it, one after the other. First, there are the "aerobic" organisms, so called because they cannot live without the presence of oxygen, which they obtain from the air. Following them, there is the second set, able to live either with or without oxygen; and these M. Bordas, in his thesis on "Putrefaction," has called "amphibious." These produce carbonic acid, also hydrogen and hydro-carbons. Lastly, there comes another category of micro-organisms, the "anaerobic" class, which do not live in oxygen, and which produce hydrogen, nitrogen, and more or less compound ammonias. These organisms follow one another, for the reason that each class secretes a poison in the presence of which it is unable to live. It then disappears, and is replaced by other colonies, and so on, until the destruction of the body is complete. This explains why it is that air is necessary to render putrefaction possible; the *first* set of micro-organisms can only exist and set up their characteristic effects when there is a certain amount of free oxygen, and this they have to obtain from the atmosphere. If this be shut off, putrefaction can be prevented for a very long time. It illustrates, also, the beautiful provision of nature; the method employed to disintegrate the body and return it to its elements as speedily as possible.

Putrefaction takes place at a different rate and in a different manner, according to the medium in which the body is placed. We have already seen the effects of withdrawing the body from a medium altogether, placing it *in vacuo*. If the body be in the air, it will decompose in one way, if in water in another; it will putrefy in a different manner still in the earth—and even here there is a great difference, according to the nature of the soil in which the body is placed.

"Micro-organisms can, of course, enter the body through the epidermis, but they seem to be very slow in doing so in the majority of cases. Usually putrefaction begins in the digestive tract. It is especially a function of the processes which take place in the intestines. M. Duclaux, who has paid much attention to the 'vibrios' of the intestines, has succeeded in determining the part they play in putrefaction. At death they swarm; they penetrate into the intestinal glands, which they destroy, find their way into the veins and peritoneum, and produce gases there, and secrete diastase, which liquefies the tissues. What is the consequence of this formation of gas and diastase? The quantity of gas produced is considerable, its tension is sometimes equal to that of 1½ atmospheres; it also pushes up the diaphragm to the third intercostal space, and drives the liquid contained in the deep vessels towards the periphery; that is what I have called the posthumous circulation."

The significance of this fact will be apparent when we come to a discussion of "Vampires." (See Appendix A.)

If a person dies from suffocation from carbonic acid gas, his tissues contain very little oxygen, and, in consequence, the first set of micro-organisms have great difficulty in gaining a foothold within the body. Brouardel gives a case in which a corpse was found to be in a perfect state of preservation two months after death —the man having committed suicide in this manner.

Of course, other causes influence putrefaction greatly. The state of health at the time of dying is known to be one great contributory cause. Patients dying of cancer putrefy very slowly for some reason. If there be food in the stomach, decomposition takes place more rapidly than if there be none, which is what we should expect *a priori*. If the coffin is badly closed, decomposition will be more rapid than if it is well sealed: the degree of moisture of the soil, or the reverse: whether the body be placed in a wooden or a leaden coffin—all these factors help to determine the rate and character of the subsequent putrefaction.

When bodies are retained in the air for some days, and putrefaction sets in, the body swells up from the created gas, and this has to be removed, in order to prevent tainting of the atmosphere. What, then, is done? This: holes are pricked in the bodies, and a lighted match applied to these minute orifices. Long, bluish flames start forth, like those of a blow-pipe. These remain ignited sometimes for three or four days, then the combustibility of the gas ceases. When decomposition is more advanced the gas will not take fire in this manner. This is due to the fact that the gases created during the later stages of decomposition are not combustible; but those in the earlier stages are.

During decomposition phosphoretted hydrogen is frequently formed. "Before the time when refrigerating apparatus was employed at the Morgue—that is to say, prior to 1882—phosphorescence was often noticed there, especially in warm weather, Wills-o'-the-wisp, which ran over and around the bodies. It was a very impressive spectacle." This has great significance, when we remember the frequent allusions to "corpse lights," etc.—spirits that were supposed to hover above the grave in the graveyard, and which have doubtless given rise to many a ghost story.

When a body decomposes under the ground, little blebs form all over the surface of the body; these are filled with a sort of serum and blood. The epidermis then separates in flakes. Gases are formed in large quantities, and when the tissues have been more or less liquefied by the action of micro-organisms, the flesh is ruptured, thus giving vent to these gases. It is curious to note that when a body is completely covered with animal excreta, it decomposes very slowly indeed; whereas precisely the reverse of this is what we should expect! I shall not do more than refer to this here,—leaving the more technical discussion for more strictly medical treatises.

When a body decomposes in water, many interesting changes take place. Dr. Brouardel assures us that "the first green patch which appears does not show itself in the region of the caecum, as it does when the body putrefies in the open air, but over the sternum;" and he adds, "I cannot explain to you the cause of this variation." Hofman calculated that putrefaction is twice as rapid in air as in water. The water in which the body is floating penetrates the periphery, and enters into the blood stream, thus preventing coagulation to anything like the extent that would take place in the air. But when the body is withdrawn from the water, putrefaction takes place with extreme rapidity.

The best account of what takes place in bodies thrown into water is the following, which I take from *Death and Sudden Death*, p. 83:—

"Bodies more frequently undergo transformation into fatty matter in the water than in the open air; this transformation is sometimes complete by the end of five or six months. If it had remained exposed in the open air, the corpse might have putrefied before so long a time had elapsed; if it had been placed in the earth, it would be necessary to take into consideration the state of the coffin and of the soil—putrefaction might be hastened or retarded thereby. In the water the phenomena of putrefaction follow the same evolutionary course as those of fermentation within the intestines. The Fenayrou case affords a demonstration of this. A druggist named Aubert was murdered in the country by a husband and wife of the name of Fenayrou, assisted by their brother. To get rid of the corpse they threw it into the Seine, after having enclosed it in a piece of lead pipe. They hoped that thus it would stay at the bottom of the water. Three days afterwards Aubert floated, though still enclosed in the lead pipe. An enormous quantity of lead would be necessary to prevent a body from rising to the surface; the only means of keeping the body at the bottom would be to open the abdomen and perforate the intestines; in this way the gases would escape as soon as they were produced."

Saponification.—This occurs only in bodies lying in water, or in very damp ground. As a general rule, this adipocere forms the more readily the fatter the bodies are. In recently dead bodies it is a white matter, soft, brittle, and somewhat watery. When exposed to the air it dries up. Bodies exposed to a hot, dry air, on the contrary, tend to "mummify," as we shall see later.

Saponification is the scientific term. It consists of the fatty matter combined at first with the ammonia disengaged by decomposition. It thus forms an ammoniacal soap. If it is in water, the lime of the water drives off the ammonia, and thus forms a lime soap, and may remain unchanged for a long period. Some think that the body may be completely saponified in a year. Bodies of infants may saponify in six weeks to thirteen months. It is probable that it begins in three or four months in water. In one case where the body was half out of the water, after fourteen months the lower part was saponified and the rest not. This soapy matter becomes ultimately broken up and washed away, if in water.

The different organs of the body decompose at very different rates, and in different manners. The bones, of course, last longest of all, becoming more and more light as time goes on, and they gradually lose their animal matter. It is asserted that the uterus is the last organ to decompose. In adults the brain decomposes slowly, in children more quickly. The liver becomes light after death, and will float when thrown into water. This is due to the formation of gas within its structure. The lungs of an adult (and those of a child) decompose in a different manner from those of a babe who has never breathed. The eye decomposes and vanishes at the end of about two months; the nails become loose about the twentieth day.

Bodies decompose at different rates. Some of them disintegrate and liquefy very speedily; others take months and even years to reach the same advanced stage of putrefaction. The causes of these differences are not known, but it would not be a difficult matter to conjecture, at least, in the majority of instances. That remarkable cases of the kind *exist,* there can be no doubt. Brouardel mentions one in which a leaden coffin was opened at the end of three months, and the corpse "looked as if it were in a bath of sweat; it was covered with moisture, and the skin was corrugated." In another case, "a woman poisoned by Pel was found, four years after death, in the exact condition in which she was when put into her coffin." In yet another remarkable case, a number of soldiers were buried together. Five years later they were disinterred, when we find that "some of them were skeletons, clothed with remains of their belts, etc., others were still in such a state of preservation that their features could be recognized." "The fact," he adds, "cannot be explained at present. All sorts of hypotheses are possible. We may assume that all these men had not the same species of microorganism in their digestive tubes" (p. 98).

As the net result of our inquiry, therefore, we find that every test of death is unreliable, with the single exception of *putrefaction.* Even here, certain discolorations and spots may appear on the surface of the body on occasion, which may be mistaken for decomposition; and it would be well to wait until unmistakable signs develop. But, on the whole, decomposition may be considered a fairly reliable test. It is, at all events, the *only* fairly reliable sign, and certainly the only sign that the layman can trust. Sir Benjamin Ward Richardson, writing on this subject, stated that only in a *combination* of signs, all appearing together, is safety to be found. He enumerates the following indications:—Respiratory failure, cardiac failure, absence of turgescence or filling of the veins on making pressure between them and the heart, reduction of the temperature of the body, *rigor mortis,* coagulation of the blood, putrefactive decomposition, absence of red colour in semitransparent parts under the influence of a powerful stream of light, absence of muscular contraction under the stimulus of galvanism, heat, or of puncture, absence of red blush on the skin after subcutaneous injection of ammonia, absence of signs of rust or oxidation of a bright steel blade after plunging it deep into the tissues. Sir Benjamin sums up the matter thus:—

"If *all* these signs point to death . . . the evidence may be considered conclusive that death is absolute. If these leave any sign for doubt, or even if they leave no doubt, one further point of practice should be carried out. The body should be kept in a room, the temperature of which has been raised to a heat of 84° F., with moisture diffused through the air, and in this warm and moist atmosphere it should remain until distinct indications of putrefactive decomposition have set in."

CHAPTER V

TRANCE, CATALEPSY, SUSPENDED ANIMATION, ETC.

Dr. George Moore, in his *Use of the Body in Relation to the Mind,* p. 31, says:—"A state of the body is certainly sometimes produced (in man) which is nearly analogous to the torpor of the lower animals— *a condition utterly inexplicable by any principle taught in the schools."* This was written years ago, but it still holds good. Very little, indeed, is known about this subject—more than some of its mere phenomena—which were recognized and carefully studied by Braid, who wrote his memoir, *Observations on Trance; or, Human Hibernation,* in 1850. When we come to inquire into the cause, the real *essence* of trance and kindred states, we find an amazing lack of knowledge on these subjects—mostly due to the fact, no doubt, that it has always been considered a mark of "superstition" to investigate such cases; and so, until the last few years, these peculiar conditions have been left strictly alone by the medical profession. When a condition of catalepsy could be shown to be due to a disordered, nervous condition, *then* it was legitimate to study such a case; but when the causes of the trance were psychological or unknown, then it immediately became "superstition"! Even today this state f affairs is not outgrown. I doubt if more than one physician in a hundred would be willing to recognize the "medium trance," e. g., as a separate state requiring prolonged psychological investigation.

In spite of the fact that Professor William James pointed out the absurdity of this attitude, it is still the one all but universally maintained.

Writing on trance, ecstasy, catalepsy, and kindred states in Pepper's *System of Medicine,* vol. v, pp. 314–52, Dr. Charles K. Mills thus defines these conditions:—

"*Catalepsy* is a functional nervous disease characterized by conditions of perverted consciousness, diminished sensibility, and especially by muscular rigidity or immobility, which is independent of the will, and in consequence of which the whole body, the limbs, or the parts affected, remain in any position or attitude in which they may be placed."

The following is the author's definition of ecstasy:—

"*Ecstasy* is a derangement of the nervous system, characterized by an exalted visionary state, absence of volition, insensibility to surroundings, a radiant expression, and immobility in statuesque positions. Commonly, ecstasy and catalepsy, or ecstasy and hysterio-epilepsy, or all three of these disorders, alternate, co-exist, or occur at intervals in the same individual. Occasionally, however, the ecstatic seizure is the only one that attracts attention. Usually, in ecstasy, the concentration of mind and the visionary appearance have reference to religious or spiritual subjects.

"*Trance* may be defined as a derangement of the nervous system, characterized by general muscular immobility, complete mental inertia, and insensibility to surroundings. The condition of a patient in a state of trance has been frequently and not inaptly compared to that of a hibernating animal. Trance may last for minutes, hours, days, weeks, or even months. In trance, as in ecstasy, the patient may remain motionless and apparently unconscious of all surroundings; but in the former or visionary state, the radiant expression and the statuesque positions are not necessarily present. In trance, as stated by Wilks, the patients may lie like an animal hibernating for days together; without eating or drinking, and apparently insensible to all objects around them. In ecstasy, the mind, under certain limitations, is active; it is concentrated upon some object of interest, admiration, or adoration. Conditions of trance, as a rule, last longer than those of ecstasy."

Braid's theory of trance is that it is:—

"A functional disease of the nervous system in which the cerebral activity is concentrated in some limited region of the brain, with suspension of the activity of the rest of the brain, and consequent loss of volition. Like other functional nervous diseases, it may be induced either physically or psychically—that is, by the influences that act on the nervous system or on the mind; more frequently the latter, sometimes both combined."

Dana [1] reported about fifty cases of prolonged morbid somnolence. He did not include among them cases of drowsiness due to old age, diseased blood-vessels, cerebral mal-nutrition, or inflammation, various toxaemïae as malaria, uraemia, syphilis, etc., dyspepsia, diabetes, obesity, insolation, cerebral anaemia, and hyperaemia, cerebral tumours and cranial injuries, exhausting diseases, and the sleeping-sickness of Africa.

He found that the prolonged somnolence showed itself in very different ways. Sometimes the patient suffers from simply a great prolongation of natural sleep; sometimes from a constant, persistent drowsiness, which he is often obliged to yield to; sometimes from frequent brief attacks of somnolence; not being drowsy in the intermission; sometimes from single or repeated prolonged lethargic attacks; finally, sometimes from period-

[1] "Morbid Drowsiness and Somnolence."—*Journal of Nervous and Mental Diseases*, vol. xi., April 18, 1884.

ical attacks of profound somnolence or lethargy which last for days, weeks, or months. He says that most cases of functional morbid somnolence are closely related to the epileptic or hysterical diathesis; but a class of cases is met with in which no history or evidence of epilepsy or hysteria can be adduced, and, though they may be called epileptoid or hysteroid, these designations are simply makeshifts; the patients seem to be the victims of a special morbid hypnosis.

Very much the same ground is taken by Dr. William B. Hammond in his *Spiritualism and Allied Causes and Conditions of Nervous Derangement,* and by Dr. Marvin in his *Philosophy of Spritualism.* It will be seen at once, the attitude assumed towards trance, ecstasy, and all kindred states in the attitude of pure materialism. Doubtless this attitude would be perfectly justifiable were it capable of covering and explaining *all the facts;* but it can fairly be said that such an interpretation of the states noted is quite incapable of explaining them all. The medium-trance is totally different from any of the states that have been discussed; it shows no identity with any of them. It is not dependent upon any morbid state of body, and cannot be regarded as a morbid symptom. Indeed, when a medium is ill, trance is generally impossible! Further, the supporters of such a view would have to account for the supernormal knowledge displayed by the medium while in the trance state. That is the *crux.* I do not care what theory of the nature of the trance state may be held, provided that it is capable of explaining all the facts. The current materialistic theories certainly cannot do this.

So little is known of this state we call *trance,* indeed,

that it has been found difficult even to define. Dr. J. Brindley James says of this condition:—

"What, then, is trance? It is a sleep-like condition that comes on spontaneously, quite apart from any gross lesion of the brain or from any toxic agency, and from which the sleeper cannot be roused even by the most energetic measures." [1]

It will be obvious that this does little more than define the state, which is as much as any work on the subject has so far attempted. Dr. James points out that, owing to our ignorance of the nature of trance and of its limitations, it is quite possible to mistake it for death on occasion unless the most exacting tests be employed. Various persons are apt to fall into this trance-like condition—"mostly educated persons of nervous temperament." [2] This trance-like condition is said to result most commonly from the following diseases or their complications:—

"Catalepsy, hysteria, chorea, hypnotism, somnambulism, neurasthenia, stroke by lightning, sun-stroke, anaesthesia from chloroform, etc., eclampsic coma in pregnancy, still-birth; cold, asphyxia from various gases, vapours, and smoke; narcotism from opium and other agents; convulsive maladies, drowning, nervous shock from gunshot, electricity, and other injuries; smothering under snow, earth, grain, or in bed; strangulation, epilepsy, mental and physical exhaustion, syncope, extreme heat and cold, alcoholic intoxication, haemorrhages, suspended animation from mental disorders, excessive emotion, fright, intense excitement, etc.; apoplectic seizures, so-called 'heart failures,' and many other diseases." [3]

1 *Trance: its Various Aspects and Possible Results*, pp. 3, 4.

2 *How the State can Prevent Premature Burial.* By James R. Williamson.

3 *Plan for Forming Associations for the Prevention of Premature Burial, etc.*, pp. 5, 6.

The condition known to us as trance is both uncertain and fluctuating. There can be no doubt that hypnotic trance, or trance induced by the mesmeric process (if there be any difference between them), is remarkably deep—so deep indeed, that Dr. Esdaile was enabled to perform, under its influence, some 261 operations of a painful and critical character, which he enumerates in his *Clairvoyance,* pp. 168–9. Such operations as the removal of a cancer of the eyeball; amputation of a thigh, a leg, an arm: various operations for the removal of tumours—operations that certainly cannot be performed easily or upon a patient who is not under the influence of an anaesthetic, mental or physical. It is amusing, in the light of our present knowledge, to see the attempts of many medical men of that day to account for Esdaile's cases. They went so far as to assert that the patients operated upon were merely hardened rogues paid to withstand the pain! The phenomenon of trance, both natural and induced, is now acknowledged, however, and recognized by all psychologists.

When we come to consider the *nature* and causes of trance, we find the greatest difficulty in forming any conception of it. All purely physiological explanations must certainly be abandoned. They do not account for the hypnotic phenomena, far less for trances of spontaneous or mediumistic type. Trance differs essentially also from sleep, though of course the two have something in common. A nearer analogy, probably, is the hypnotic trance; and it has occurred to me that the mediumistic trance might be a type of hypnotic influence *from "the other side,"* just as the hypnotic trance that we know is a species of mental influence from *this* side. In other words, both hypnotic and mediumistic trances may be samples of mental influence—the one from the mind of a living, the other from that of a dead operator. This would seem to be strengthened by the fact that mediums are frequently very insusceptible to hypnotic or even to normal suggestion from operators on this side. Mrs. Piper has been tested for this, for example, and no trace of any faculty of thought transference has been found, and only a light state of hypnosis could be induced in her, even after prolonged attempts. This would seem to indicate that the more an individual spirit is *en rapport* with another world, the less is it *en rapport* with this. Mr. Myers, in his *Human Personality,* has distinguished three distinct types of Trance. He says:—

"The first step, apparently, is the abeyance of the supraliminal self, and the dominance of the sub-liminal self, which may lead in some cases to a form of trance (or what we have hitherto called secondary personality), where the whole body of the automatist is controlled by his own sub-liminal self, or incarnate spirit, but where there is no indication of discarnate spirits. The next form of trance is where the incarnate spirit, whether or not maintaining control of the whole body, makes excursions into, or holds telepathic intercourse with, the spiritual world. And, lastly, there is the trance of possession by another, a discarnate spirit. We cannot, of course, always distinguish between these three main types of trance, which, as we shall see later, themselves admit of different degrees and varieties."

Mr. Myers contends elsewhere that the simplest aspect of trance is "suggested sleep," which would seem to agree somewhat with the theory advanced above. *Dreams,* the author shows, by analogy, to be "bubbles breaking upon the surface from the deep below." Extending his analogy, he has conceived clairvoyance as a state in which the spirit of the seer is enabled to leave the body and travel through different scenes and localities. In *ecstasy,* the soul would change its environment and pass for a time into the spiritual world, retaining such relations to the organism as enables it to return to its ordinary condition. And so, our author goes on, "when the last change comes, and we ask ourselves with what added ground for speculation we now strain our gaze beyond that obscurest crisis," we find—that death is an irrevocable self-projection of the spirit; that condition in which the spirit has emerged from the body, and, because of altered physical conditions, is unable to return to it.

Sleep and Death

Many analogies have been drawn between sleep and death, and death is often called "the last sleep." But there is always this distinction between the two, that in the one case we revive and return to animate the body, and in the other case we do not. Where consciousness is, what becomes of it during the hours of sleep, has always been one of the most bitterly disputed points in psychology. Certain it is that self-consciousness is absent *pro tem.;* but whether it is annihilated, as materialism teaches, or merely withdrawn, as the opposite school avers, is a question that is as far as ever from being satisfactorily answered. Many are the battles that have been fought over this point, but none of them has ever been won! Truly the field is open, and the world is at the feet of the man who shall discover the innermost nature of sleep. It is equally a mystery with death, and it is probable that there is some close interdependence between them. Veridical and supernormal dreams; cross-correspondence between dreams and the statements made by trance mediums, etc., would seem to indicate that the human spirit is simply withdrawn during the hours of sleep—being revivified in some other sphere. However, these are questions into which we cannot enter now.

Both trance and catalepsy occur spontaneously; both may also be induced artificially by hypnotism. Both are mistaken for death, and in many respects they are very similar. In catalepsy the body is rigid, whereas in trance this is very rarely the case—this forming the chief mark of distinction (external indication) between the two states. What the *internal* differences are we do not know. Various attempts, however, have been made to define them. Dr. Franz Hartmann, e. g., thus distinguishes them:—There seems hardly any limit to the time during which a person may remain in a trance; but catalepsy is due to some obstruction in the organic mechanism of the body on account of its exhausted nervous power. In the last case the activity of life begins again as soon as the impediment is removed or the nervous energy has recuperated its strength.

Whatever the innermost nature of this trance state may be, it seems certain that some individuals have the faculty of inducing this condition at will, just as it may be induced by hypnotic processes from without. The Yogis and Fakirs of India doubtless possess this power to some extent.[1] Braid gave what was probably the first authentic account of their remarkable cases of suspended animation and voluntary interment; which are also to be found narrated in more recent works—e. g.,

[1] See my book *Higher Psychical Development* for an account of the methods employed in order to render this possible.

Hudson's *Law of Psychic Phenomena*, pp. 309–20. There are many such cases, and it is reported that a number of persons have been buried alive in consequence of the inability of the attending physician to distinguish the induced state from true death. It is not to be wondered at; and until these states and conditions receive the study and attention they deserve, such cases of premature interment will probably continue to occur.

When we come to inquire into the immediate causes of catalepsy and allied states, we find that very little is known about these conditions. Dr. Alexander Wilder, in his *Perils of Premature Burial*, p. 19, says that:—

"We exhaust our energies by overwork, by excitement, too much fatigue of the brain, the use of tobacco, and sedatives and anaesthetics, and by habits and practices which hasten the Three Sisters in spinning the fatal thread. Apoplexy, palsy, epilepsy, are likely to prostrate any of us at any moment; and catalepsy, perhaps, is not far from any of us."

Again, Dr. W. R. Gowers, in Quain's *Dictionary of Medicine*, p. 216, says:—

"Nervous exhaustion is the common predisponent; and emotional disturbances, especially religious excitement, or sudden alarm, and blows on the head and back, are frequent immediate causes."

Dr. James Curry, on the other hand, thinks that fainting fits and losses of blood are the chief factors in inducing these death-counterfeits. (See *Observations on Apparent Death*, pp. 81, 82.) M. Charles Londe, in *La Mort Apparente*, p. 16, says:—

"Intense cold, coincident with privations and fatigue, will produce all the phenomena of apparent death. . . . "[1]

[1] I have considered freezing to death on pp. 62–64.

Struve, in his essay on *Suspended Animation,* p. 140, takes the same view. It has frequently been pointed out that the sequelae of certain diseases, the use of narcotics, etc., will result in states that cannot be distinguished from death. These cases of suspended animation will sometimes last for many days, as has frequently been shown; and if the body be buried during this interval, we should have a case of "premature burial."

How long may a body cease to show signs of life, and yet be revived? That is a much disputed point; but there can be no question that, if air be permitted access to the body of the patient, it can be revived after a very long period—a period not of hours, but of days and weeks. Indeed, Sir Benjamin Ward Richardson said on this point:—

"We are at this moment ignorant of the time when vitality ceases to act upon matter that has been vitalized. Presuming that an organism can be arrested in its living in such manner that its parts shall not be injured to the extent of actual destruction of tissue, or change of organic form, the vital wave seems ever ready to pour into the body again so soon as the conditions for its action are re-established. Thus, in some of my experiments for suspending the conditions essential for the visible manifestations of life in cold-blooded animals, I have succeeded in re-establishing the condition under which the vital vibrations will influence, after a lapse not of hours, but even of days; and for my part I know no limitation to such re-manifestation, except from the simple ignorance of us who inquire into the subject." [1]

Assuredly this is a significant admission! In the light of this fact, certain historic cases of "raising the dead" might be re-interpreted, and put upon a rational basis.

[1] *Ministry of Health,* pp. 154–5.

There can be no doubt that re-animation has taken place after very long intervals on occasion—even when there has been no external sign of life in the interval. Of course the time would be comparatively brief, if the supply of air were cut off, or the heart actually stopped. In a coffin of the usual dimensions, it has been estimated that from twenty minutes to an hour would insure death from suffocation. But even here we must allow, as Tebb and Vollum point out (*Premature Burial,* p. 211), for a certain persistence of the vital energy, which continues after all atmospheric air has been cut off.

"Experiments on dogs show that the average duration of the respiratory movements after the animal has been deprived of air is four minutes five seconds. The duration of the heart's action is seven minutes, eleven seconds. The average of the heart's action after the animal has ceased to make respiratory efforts is three minutes fifteen seconds. These experiments further showed that a dog may be deprived of air during three minutes fifty seconds and afterwards recover without the application of artificial means." [1]

It may be said that with modern improvements, and with the aid of artificial stimulants, this period has been very greatly exceeded.

In this connection I may mention certain facts of interest that are to be noted in the animal world. Professor S. J. Holmes, writing in the *Popular Science Monthly,* for February, 1908, calls attention to the instinct of feigning death among various animals and insects. Some of them assume attitudes that render them almost indistinguishable from their surroundings; others draw themselves up into a ball; still others remain in

[1] Report on "Suspended Animation," by a Committee of the Royal Med. Chirur. Society, July 12, 1862.

a state of apparent catalepsy, in whatever attitude they are placed, this state lasting for an hour or even longer. It is interesting to note in this connection that the attitudes assumed by these various animals at such times often bear no resemblance to the attitudes they assume in death. Darwin observed this and said:—

> "I carefully noted the simulated positions of seventeen different kinds of insects belonging to different genera, both poor and first-rate shammers. Afterwards I procured naturally dead specimens of some of these insects (including an Iulus, spider, and Oniscus) belonging to distinct genera, others I killed with camphor by an easy slow death; the result was that in no instance was the attitude exactly the same, and in several instances the attitudes of the feigners and of the really dead were as unlike as they could possibly be."

Professor Holmes does not consider that in the insects at least, this feigning of death is a conscious impulse, but rather of the nature of a reflex action. He states that the mere handling or touching of certain insects—for example, the water scorpion—will cause them to feign death for an hour, even if they are left entirely alone, or covered up, and their tormentor leaves the room. It is interesting to note, also, that these creatures cannot be made to feign death by any amount of handling under water. As soon as they are in the air, however, they feign death repeatedly. As soon as the state has worn off, if they are touched again, they again feign death for an hour or so, and this may be repeated a number of times in succession.

Among the higher animals, on the contrary, such as the fox, it would appear that this instinct is largely an act of consciousness, and that they are perfectly aware of their surroundings, and of the reason for their feigning in this manner. A fox, when feigning death, will often cautiously open its eyes, raise its head, look around, and finally scamper off, if its pursuers have withdrawn to a safe distance.

It would appear that, in the majority of cases, especially among the insects, the induced state resembles that of catalepsy; the muscular rigidity noticed—which is intense—would indicate this, and the fact that they suffer a great amount of maltreatment (pricking, mutilation, burning, etc.) without showing any signs of sensibility, would seem to show that this is lost, and that more or less complete anaesthesia is present. The state is probably closely akin to what has been called "hypnotism" in the lower animals. Practically nothing is known of that condition of the nervous system which makes such results possible, and this is as true of the higher as of the lower creatures.

CHAPTER VI

PREMATURE BURIAL

1. Cases

We have seen, as the result of the two preceding chapters, that there is no certain sign of death (with the single exception of putrefaction, which is not generally waited for), and that there are, on the contrary, many states and conditions which very closely simulate death; that, for days in fact, it is almost impossible to distinguish true from false death—so similar are they. The question here arises: Is it not possible, and in fact probable, that in certain cases a living person has been buried by mistake, under the impression that he is dead? Might it not be quite possible that accidents of the sort occur and premature burial take place? It would certainly seem that such *must* be the case; and when we turn to an account of the actual facts we find that such *has* happened very frequently. It is improbable that premature burial takes place as frequently as it did some years ago, but it is doubtless true that many cases are on record, amply testifying to the fact that it has occurred with horrible frequency, from time to time, in the past. A large number of such cases, authenticated more or less fully, are to be found in Tebb and Vollum's *Premature Burial,* in Franz Hartmann's *Buried Alive,* and in the *Encyclopedia of Death,* vol. ii, pp. 7-114. A great mass of cases are here adduced; and, although Dr. David Walsh attacked the evidence in his little book, *Premature Burial,* there can be no doubt that a large number of the cases printed stand the test of scrutiny, and are veritable cases of "premature burial." Similar cases are continually coming to the attention of the public from time to time and it is surely high time that some means be adopted to check this evil. It is true that there is a Society for the Prevention of Premature Burial—both in England and America—but it is unable to accomplish much, owing to the tyranny that it has encountered in more than one direction. Such a movement deserves the whole-hearted support of the people; and I shall now endeavour to lay before the reader my reason for taking this stand so strongly.

Nothing that the human mind can conceive can appeal to the imagination as more horrible than the idea of premature burial. To awake in a coffin—cold, dark, and helpless—far beneath the surface of the ground, and know that the living tomb is one from which it is impossible to escape, suggests a tragedy that is in every sense appalling. If we attempt to picture such a fate, it is easy to comprehend how the agony of a whole lifetime may be compressed within the few minutes that elapse between the moment when the victim awakes to the horror of his position and the time when he again lapses into unconsciousness, as the effect of suffocation. It is not strange that such a subject should have appealed to the writer of realistic fiction, but we must not imagine that these cases occur only in the pages of the sensational novel. In writing upon this subject, Professor R. L. O. Roehrig, formerly of Cornell University, said:—

"The possibility of premature burial always exists, for that there is real danger of being buried, embalmed, dissected, or cremated alive has been fully acknowledged by various unquestionable, highly respectable authorities, and many celebrated authors have written on this particularly important subject, among them Alexander Humboldt, Hartmann, and Hufeland. All have shown that in every case of death which cannot be plainly accounted for by violent external causes, fatal vulnerations, accidents by firearms or other deadly weapons, suicide or murder, it is of the utmost importance to abstain from all sudden alarm and meddlesome interference, and most patiently to wait until every possible doubt as to the real and entire extinction of life has been absolutely removed.

"Under no conditions should the fear of ridicule, supercilious contempt, or mockery coming from the thoughtless, or any other sort of intimidation, influence us in our conduct on so grave an occasion. Nobody can be certain that he will not at some time have to undergo this horrible misfortune, for the most celebrated and experienced physicians have been misled by appearances, while even the assertions of the public inspectors of the dead have often led to the most deplorable consequences."

Prone as the scientist may be to question the accuracy of the assertion that, at the smallest average, one person is buried alive in the United States every twenty-four hours, it is important to note that the London Humane Sociey has reported the fact of having brought back to life no less than 2175 apparently dead persons within a term of twenty-two years; that a similar society in Amsterdam restored 990 persons in twenty-five years; and that the Hamburg Society saved 107 persons from premature burial in less than five years. Personally, I know of several cases of this kind, and, in one instance a prominent New York physician recently discovered to his horror that the body he was dissecting was that of a live man. Professor Roehrig, who asserts that he has saved many persons from this fate, states that he once rescued a child from the dissecting table, in spite of the insulting mockery of all the other physicians who were in attendance. In view of these facts, it is not difficult to believe that the following gruesome experience, related by a French physician in the Paris *Figaro,* may be fact, not fiction:—

"Five years ago," he writes, "I was preparing for an examination, and went one night alone into the dissecting room for the purpose of studying certain abdominal viscera, carrying a light in my hand. An insane woman, having died on the day before, was extended naked upon the marble slab. I placed my candle upon her chest, and made a cut through the skin over the stomach. At that moment the supposed corpse gave a terrible scream, and, rising up, caused the light to fall and become extinguished. Then a terrible struggle began; the woman, with one of her cold clammy hands took hold of my hair, and with the other clawed my face with her finger nails. I was beside myself with terror, and blindly struck about me with the scalpel which I still held in my hand. Suddenly my knife struck an obstacle; a sigh followed, the grasp on my hair was loosened, I fainted, and knew nothing more. When I awoke it was daylight; I found myself upon the floor lying beside the bloody corpse of the woman whom I had killed, as my knife had gone directly to her heart. I replaced the corpse upon the table and said nothing about it; but the recollection of this event fills me with horror, while the marks which the nails left upon my face are still there."

It has been pretty authoritatively asserted that Mlle. Rachel, the celebrated actress, was embalmed while still alive, and there are those who will always believe that Washington Irving Bishop, the distinguished mind-reader, died from the effects of an autopsy performed while the unfortunate man was in one of the trances to which he was frequently subject. It is also stated that the mother of the famous General Lee was buried alive and resuscitated two years before his birth. Although pronounced to be dead by her physician, she regained consciousness sufficiently during the process of interment to attract the attention of the sexton. Ebenezer Erskine, one of the founders of the (United) Free Church of Scotland, is also said to have been born after the burial of his mother. As in the case of Mrs. Lee, Mrs. Erskine was buried while in a trance. As the gravedigger had noticed that there was a valuable ring on one of her fingers he determined to secure it, and, stealing to the new-made grave during the night, he removed and opened the casket, and cut off the finger on which the ring had been placed. It was by this act of felony that her life was saved.

A comparatively short time ago, George Hefdecker, a farmer living near Erie, Pa., died suddenly of what was supposed to be heart failure. The body was buried temporarily in a neighbour's lot in the Erie Cemetery, and when, some time later, the transfer to a newly-purchased family lot was made, the casket was opened at the request of the relatives. To their horror it was then discovered that the body had turned completely round, and the face, as well as the interior of the coffin, bore unmistakable traces of the terrible struggle that had occurred.

A similar story comes from Petrograd, Russia, in connection with the interment at Tioobayn, near that city, of a peasant girl named Antonova. She had presumably died, and was buried, but after the gravedigger had completed his work he was startled by sounds that seemed to come from the new-made grave. Instead of removing the coffin and breaking it open, however, he rushed off to find a doctor, and when he and the public officials arrived it was too late. The casket contained a corpse, but, as the position of the body clearly proved, death had only just taken place.

When the question of premature burial came up for discussion before the French Senate some years ago, a most remarkable story was told under oath by Cardinal Archbishop Donnet. In part, his testimony was as follows:

"In the summer of 1826, on the close of a summer day, in a church which was exceedingly crowded, a young priest who was in the act of preaching was suddenly seized with giddiness in the pulpit. The words he was uttering became indistinct; he soon lost the power of speech, and sank down on the floor. He was taken out of the church and carried home. All was thought to be over. Some hours after the funeral bell was tolled, and the usual preparations made for the interment. His eyesight was gone; but if he could see nothing, he could hear, and I need not say what reached his ears was not calculated to reassure him. The doctor came, examined him, and pronounced him dead; and after the usual inquiries as to his age, the place of his birth, etc., gave permission for his interment the next morning. The venerable bishop, in whose cathedral the young man was preaching when he was seized with the fit, came to the bedside to recite the 'De Profundis.' The body was measured for the coffin. Night came on, and you can easily feel how inexpressible was the anguish of the living being in such a situation. At last, amid the voices murmuring around him, he distinguished that of one whom he had known from infancy. That voice produced a marvellous effect, and he made a superhuman effort. Of what followed I need only say that the seemingly dead man stood next day in the same pulpit. That young priest, gentlemen, is the same man who is now speaking before you, and who, more than forty years after that event, implores those in authority not merely to watch vigilantly over the careful execution of the legal prescriptions with regard to interments, but to enact fresh ones in order to prevent the occurrence of irreparable misfortunes."

Bouchut, in his *Les Signes de la Mort,* p. 43, gives the following case:—

"A person of high standing was taken with one of those diseases in which death usually does not occur suddenly, but is preceded by certain signs. The physician who attended him found him one evening in a dangerous state, and when he visited him again the following morning, he was told upon entering the house that the patient had died during the night. They had the body already placed in the coffin, but the doctor, doubting that death could occur so suddenly, caused the supposed 'dead' to be put back into bed. The man soon revived, and lived for many years afterward."

Dr. Hartmann gives the two following cases collected by himself, and published in his *Buried Alive,* pp. 52–53:—

"At Wels (Austria) a woman died, and as no signs of putrefaction appeared at the end of five days, all sorts of means were resorted to, to revive the body. They were of no avail, and it was finally resolved not to delay the burying any longer. On the night preceding the funeral a large crowd met for the purpose of holding the 'Wake,' It was a merry party, and some of those present got drunk and amused themselves in making jests with the corpse and offering it liquor. In the midst of the merry-making, the woman woke and sat up in her coffin! The company ran away, and when they returned they found that the woman had gone to bed, where she slept, and was well the next day. She had been conscious of all that had taken place, but had not been able to move.

"In another town in Austria, a student made a bet that he would not be afraid to go at night to the graveyard, open a grave, steal the corpse, and carry it to his room. This he did accordingly, and the grave he opened happened to be that of a young girl who had been buried on the previous day. He took the body upon his shoulders and carried it to his room, where he put it upon a lounge near the stove. He then went to sleep. During the night he was awakened by a noise. The girl had wakened from her trance, and was sitting up. He was so much terrified that his hair turned white; but the girl, thus saved, returned to her parents."

Sometimes the termination of such cases is not so fortunate, however. It will be observed that in the following case, reported in the *British Medical Journal,* April 26, 1884, p. 844, death resulted from the interment:—

"The *Times of India,* for March 21, has the following story: 'On last Friday morning the father of a large family at Goa, named Manuel, aged seventy years, who had been for the last four months suffering from dysentery, appearing to be dead, preparations were made for the burial. He was placed in a coffin and taken from his house at Worlee to a chapel at Lower Mahim, preparatory to burial. The priest, on putting his hand on the man's chest, found his heart still beating. He was thereupon removed to the Jamsetjee Jejeebhoy Hospital, where he remained in an unconscious state up to a late hour on last Friday night, when he died."

The following case is quoted in Tebb and Vollum's *Premature Burial,* p. 55:—

"A young married woman residing at Salon died shortly after her confinement in August last. The medical man, who was hastily summoned when her illness assumed a dangerous form, certified her death, and recommended immediate burial in consequence of the intense heat then prevailing, and six hours afterwards the body was interred. A few days since, the husband having resolved to re-marry, the mother of his late wife desired to have her daughter's remains removed to her native town, Marseilles. When the vault was opened, a horrible sight presented itself. The corpse lay in the middle of the vault, with dishevelled hair, and linen torn to pieces. It evidently had been gnawed by the unfortunate victim. The shock which the dreadful spectacle caused to the mother has been so great that fears are entertained for her reason, if not for her life."

Another remarkable case is the following (*Encyclopedia of Death*, vol. ii, p. 107):—

"Thirty-four years ago, a man by the name of John Hurelle was pronounced dead by three doctors, who held an examination. Everything was prepared for the funeral; the guests were invited, a clergyman summoned, and the body placed in a coffin. On the morning when the funeral was to occur, the mother thought she saw signs of life, though four days had passed since he was said to have been dead. The funeral did not take place. When those present took the seemingly lifeless body and placed it on a bed, the man said: 'Let me'— and then stopped. For eight months he lay in a sort of stupor, while his mother gave him nourishment. At the expiration of that time he regained consciousness, and finished the sentence by saying 'be.'"

Another case collected by Dr. Hartmann himself, is the following:—

"In a small town in Prussia, an undertaker, living within the limits of the cemetery, heard during the night cries proceeding from within a grave in which a person had been buried on the previous day. Not daring to interfere without permission, he went to the police and reported the matter. When, after a great deal of delay, the required formalities were fulfilled and permission granted to open the grave, it was found that the man had been buried alive; but he was now dead. His body, which had been cold at the time of the burial, was now warm and bleeding from many wounds, where he had skinned his hands and head in his struggles to free himself before suffocation made an end to his misery."

"In the month of December 1842, an inhabitant of Eyures, in France, died and was buried, a few days afterwards a rumour began to spread that his death was due to an overdose of opium having been given to him by a physician. Finally, the authorities ordered the grave to be opened, and it was found that the supposed dead man had awakened and opened with his teeth the veins of his arm for the purpose of ending his tortures, and then he had died in his coffin."—(Lenormand, *Des Inhumations Précipitées,* p. 78.)

Many persons seem to think that premature burials are few and far between. There was never a greater fallacy, says M. Tozer:—

"Some years ago the Paris *Figaro* dealt at considerable length with the subject of the possibility of premature burial occuring somewhat frequently, and within fifteen days the editor received over four hundred letters from different parts of France, all from persons who either had been almost buried alive, or who knew of such cases." [1]

Dr. Franz Hartmann, immediately after the publication of his book on the subject, and within two months (May—June, 1896), received no less than sixty-three letters from persons who had escaped premature burial through fortunate accident. When such wholesale

[1] *Premature Burial,* by Basil Tozer, p. 16.

numbers are observed, what are we to think but that premature burial, so far from being a great rarity, is a not infrequent phenomenon—happening constantly in our very midst?

In an article in the *Insurance News* for April 1, 1901, George T. Angell, founder and president of the American Humane Education Society, says that

"Nothing can be more certain than that large numbers (and perhaps multitudes) of persons have been buried alive, and that many, after having been pronounced dead, have shown signs of life in time to save themselves from such burial, and have declared that, *while unable to move they were fully conscious of what was said and done about them.* My own father barely escaped such burial, being declared by his physicians dead. There are in Boston alone many thousands of persons living in hotels and boarding-houses where, whenever death is declared, every effort will be made to send the body of the supposed deceased, at the earliest possible moment, to the undertaker, the crematory, or the grave. In not one case in a hundred will the body be permitted to remain in the hotel or boarding-house until the beginning of decay."

The following is a recent case of this character, which has the added interest that the subject happened to be a well-known "medium," and that she claimed she saw her dead husband, during the period of insensibility—thus constituting a case of a (so-called) "vision of the dying."

IRONTON, O, April 8, 1920—Mrs. Elizabeth Blake, seventy-two years old, of Coryville, a spiritualistic medium, startled relatives and friends by returning to life four hours after a physician had pronounced her dead. She is weak and is not expected to live twenty-four hours.

Mrs. Blake talked little, but she is quoted as having whispered that she had seen her husband, who died last November.

The aged spiritualist was stricken with pneumonia several days ago, and at noon the death pallor overspread the body, no sign of heartbeat could be detected, and rigor mortis set in. Relatives and friends flocked to the house, an undertaker was summoned and the newspapers were provided with an obituary.

Shortly after 1 o'clock persons in the room where the supposed corpse was lying experienced a shock upon seeing that the woman's eyes had opened and a pink tinge was suffusing her cheeks. Restorative measures were applied, and Mrs. Blake recognized those at the bedside.

Finally, the following case, recently reported from Germany,—a country where investigations of this character are made with the utmost scientific exactitude:—

BERLIN, Nov. 14, 1919—The German medical profession is deeply interested in an extraordinary case of catalepsy or asphyxia which the greatest authorities, after a careful examination, can only explain as a relapse into hibernation, based upon the theory that, like the wild beast, prehistoric man would pass the Winter months in a continuous, deathlike sleep, and that the case in question must be regarded as sudden atavism, brought on by peculiar circumstances.

About a week ago an auto party driving through the Tiergarten discovered the lifeless body of a young woman, dressed as a sick nurse, which they conveyed to the nearest hospital, where, after the usual tests, she was pronounced dead, and, with her clothes soaked in rain, placed in a coffin. Fourteen hours later some police officials called at the morgue to search for any articles by which to identify the body and notify the relatives. To their great surprise and mortification, they found the heart still beating.

They had the "living corpse" transferred to a hospital, where the young woman, under proper treatment, soon revived, and after a few days when the danger of pneumonia had passed, she returned to her parents.

The highest medical authorities investigated the charges made against the hospital physician who had signed the death certificate that he had only superficially examined the body and that the test employed was insufficient. But it was proved beyond doubt that the physician performed the investigation absolutely correctly, in the presence of several colleagues, all of whom agreed that life was extinguished. There were indications that the young woman had taken a strong dose of morphium. She had, as it developed afterward, tried to commit suicide on account of unrequited love.

After having taken the poison, the young woman wandered about the Tiergarten in the pouring cold rain until she fainted, and was picked up several hours afterward.

As Professor Dr. Ernest Rautenberg, who presided over the investigation, explained, the effect of the cold and rain created the condition of reducing respiration and circulation to an infinitesimal state, while the concurring catalepsy caused by the cold and rain made the body insensible to the otherwise deadly dose of morphium and protected her against pneumonia, which under normal conditions the girl could never have escaped. The morphium was removed from the girl's stomach in an unassimilated condition, and after a few days of careful observation she returned to her parents healthy and sadder, but much wiser.

Dr. Rautenberg believes Indian fakirs undergo very much the same process before they permit themselves to be buried under ground for many weeks.

Dr. Henry J. Garrigues, of New York City, in a paper read before the Society of Medical Jurisprudence, contended that any law permitting burial without thorough tests to determine the extinction of life was nothing short of homicidal. Under the law of "necessary precautions," he said, "there is nothing to prevent anybody from being buried alive or frozen to death in an undertaker's ice-box.'' His objection to the laws that now exist so generally throughout the country is based upon the fact that they are designed to protect the community, without regard to the protection of the person supposed to be dead. ''And yet,'' as Dr. Garrigues admitted, ''the question of whether a person is dead or alive is most difficult to decide. If the action of the vital organs is suspended, every appearance of death may be produced, when, under proper manipulation, they may be restored to life.''

In citing the counterfeits of death, Dr. Garrigues referred to persons who, though taken from the water apparently dead, were afterwards resuscitated, and he stated the belief that, if it were not so common to believe that people were dead merely because they were cold and limp, many others would be revived. Asphyxiation, heart failure, apoplexy, intoxication, lightning stroke, anaesthetics, narcotics, concussion—all these produced the counterfeits of death, and often so closely resembled it that the science and the experience of the physician were frequently at fault. Thus the danger of mistaking live persons for dead remains, even after all tests for determining death have been tried. There is not one but which may fail under certain conditions. The most common test of all, that of trying to ascertain if the breath has stopped, is the one that is usually made, and yet science knows of many cases of suspended animation where breathing has ceased for fully forty-eight hours. The same is true regarding the stopping of the heart, and so on through the entire list. There have been cases of suspended animation in which all signs have failed, and yet the patient recovered. In his opinion, the only sure indication of death is the decomposition of the body.

Dr. Garrigues' opinions upon this subject were fully upheld by Dr. John Dixwell, of Harvard University. In an address before the Committee on Legal Affairs of the Massachusetts Legislature, February 12, 1908, he stated that he personally had narrowly escaped premature burial. "During an illness, in the early seventies," he said, "very eminent physicians determined that I was dead, but I am alive today, while they are all dead. Accordingly I know that this horror exists as a fact. It is ridiculous to dispute it. I recall a case at the Massachusetts General Hospital. A woman had been sent there suffering from bronchitis. After a time it was decided that she was dead, and she was sent to the morgue. There she suddenly woke up, and is alive today."

"I have often been told," says Dr. Alexander Wilder,[1] "that the modern practice of embalming made death certain. I admit it; but those who are too poor to pay for this funeral luxury must yet take the chance in the old-fashioned way. There is no doubt, however, that the number annually put to death by embalmers is sufficiently large to demand attention. An investigator of this subject in New York has openly declared his belief that a considerable number of human beings are annually killed in America by the embalming process."

Dr. Edward P. Vollum, surgeon in the United States Army, is another physician who has written freely upon the danger of premature burial. In addition to collaborating with Tebb in compiling a book upon this subject, he is the author of several papers treating of the dangers of burial alive, from one of which I quote:—

1 *Burying Alive a Frequent Peril*, p. 12.

"Any one whose vital machinery is thrown out of gear by excesses, strains, or depressing causes may pass into and out of this transitory state if they have a reserve of strength. Shocks cause apparent death, such as from gunshot, strokes of lightning, charges of electricity, concussion, heat and sunstroke, fright, intense excitement, etc. So do exhaustions from mental and physical exertion, especially in the badly nourished, asphyxia from various causes, intense cold, anaesthesia, intoxicants, haemorrhage, narcotism, convulsive disorders, so-called heart failures and apoplectic seizures, epilepsy, and syncope.

"The above cases are quite plain, and many are saved by medical aid. But there are other forms of this mysterious state that may defy the highest medical skill and all known tests and signs. These are the constitutional cases, due to some warp of temperament, as seen in trance, catalepsy, cholera, auto-hypnotism, somnambulism, etc., which, like hibernation, are inexplicable by any principles taught by science. We know but little of these idiosyncrasies except that they are usually hereditary, and that their victims easily fall into a deathlike lethargy from overwork, worry, and foul air, and that during their attacks efforts at resuscitation should be kept up until putrefaction appears, lest they be mistaken for dead and disposed of accordingly. Quain's *Dictionary of Medicine* says: 'The duration of the trance has varied from a few hours or days to several weeks or months.' The British medical press during the last fifty years has given numerous cases which revived from the consciousness of the preparations for closing the coffin. Many notables have been subject to this disorder, such as the great anatomist Winslow, the French Cardinal Donnet, and Benjamin Disraeli. The last-named lay in this state for a week.

"All such cases are in peril because of their uncertainty. Of course, old cases of heart disease and apoplexy may be recognized by the patient's physician, but, as a rule, the diagnosis cannot be sure without an autopsy. All signs of death are deceptive, and all these cases should be held as not beyond resuscitation until decomposition appears. Hufeland says: 'Death does not come suddenly; it is a gradual process from actual life to apparent death, and from that to actual death.'

"The revivals sometimes reported during epidemics of cholera, small-pox, and yellow fever depend, as in so-called sudden deaths, upon the fact that the patients are usually struck down in their ordinary health with a reserve of strength which bridges them over after the force of the disease is spent.

"The estimates of such disasters are based upon the discoveries made when the dead are removed from cemeteries, as is done in some great cities every five years. A portion of the skeletons are always found turned to one side or on the face, twisted, or with the hands up to the head. These are counted as living burials. And then there is the admittedly large number of narrow escapes from being buried alive, recovered, as a rule, by some chance. Hidden and mixed with ignorance, laxity, and indifference as this whole matter is, the authorities naturally differ in their views as to the frequency of these cases. A personal inquiry in Europe and in the United States for several years past has convinced me that they are alarmingly frequent. The proportion of discovered cases must be small compared with those that never come to light. Dr. Lionce Lenormand, in *Des Inhumations Précipitées,* says that a one-thousandth part of the human race have been and are annually buried alive. M. le Guen, in *Dangers des Inhumations Précipitées,* estimates premature burials at two a thousand. He collected 2313 cases from reliable sources. Hundreds of foreign authorities with similar views could be given. Dr. Moore Russell Fletcher, in *One Thousand Persons Buried Alive by Their Best Friends* (Boston, 1890), gives many horrors taken from American sources. Carl Sextus, of New York, collected in eighteen years 1500 cases of death counterfeits of scientific value. He estimates living burials at two per cent."

I have now given a number of cases of premature burial, or cases in which burial would have taken place shortly had not some fortunate and unforseen accident happened. A number of similar cases will be found detailed in the authors quoted, and in other works upon the subject. Bruhier, in his work, *Dissertations sur l'Incertitude des Signes de la Mort, etc.*, produces accounts of 181 cases, among which there are those of 52 persons buried alive, 4 dissected alive, 53 that awoke in their coffins before being buried, and 72 other cases of apparent death. Hartmann himself gives more than a hundred cases. Tebb and Vollum collected an equal number, and many cases appear elsewhere in the literature upon this subject. Enough has been said at all events to show how numerous these cases are; and it becomes evident that some steps should be taken to prevent such burials from taking place. It is my hope that the publication of this book will at all events stimulate public interest in this direction, and help to initiate some widespread movement for the prevention of such horrible cases as those described.

2. Efforts to Prevent Premature Burial

During the last few years the question of the prevention of premature burial has been taken up by several of the State legislatures, and laws have been suggested, and, in some cases enacted, tending to reduce the possibility of such a catastrophe. One of the best examples of such legislation is the bill presented to the Massachusetts Legislature. This provides that local boards of health shall be notified within six hours of the death of any person, and that, as soon as possible, an examination shall be made of the reported deceased, and that certifications of death shall be issued only after ten tests have been made—for heart action, respiration, circulation, *rigor mortis,* etc., and the use of subcutaneous injections of ammonia.

This subject of premature burial, now being agitated in the United States, was thoroughly considered in Europe, beginning more than a century ago. As Dr. Vollum has shown in his article on *Final Tests for Death,* France first recognized the necessity for legal protection against these dangers. Germany was the first to put them in force. Then followed France, Austria, Belgium, Spain, the Netherlands, and Scandinavia. The pith of these laws is in the requirement of an expert examination of the apparently dead independently of the attending physician. In Germany, Austria, and Belgium the examiners, called inspectors of the dead, are officers of the State, specially qualified for their duties. In the other States mentioned they are physicians of standing, also qualified. They must decide the cause and fact of death, and register a certificate of verified death before a burial permit can be issued or the body disturbed in any way with the view of embalming, autopsy, burial, or cremation. The underlying principle of these laws is well expressed in the Austrian imperial law thus:—

> "That the only sure sign of death being general decomposition, which as a rule comes late in the case, the examiner of bodies, in the absence of this proof, must not be guided by any single sign, and must base his conclusions on an assemblage of all signs that point to death, and to any injuries that may involve the vital apparatus."

These laws, framed both in the interests of the State and the individuals, are supported by the legal and the medical professions, and have always given satisfaction to the authorities and comfort and a sense of safety to the people, excepting in France, where the period allowed before burial was only twenty-four hours, and the inspections were thought to be rather perfunctory, especially in Paris. The German and Austrian systems are alike, excepting in the former all bodies must go to the waiting mortuaries; in the latter this is voluntary, as it is in the other States named.

The German sysem is best seen in Munich. This city of about 50,000 people is divided into twenty-one burial districts, in each of which there is an inspector of the dead, with an alternate, besides the woman who makes the toilet of the body, called *Leichenfrau,* and who arranges the funeral appointments. She is also qualified by a technical examination. The attending physician is always present at the death crisis. He gives his verdict of death, but the law does not trust his unsupported opinion, however celebrated he may be. The inspector comes, and in the meantime nothing about the body must be touched by any one. He makes his certificate, which covers every possible point in the case, and this is countersigned by the attending physician. Delay and resuscitation may be employed at this stage if the inspector sees fit. Ordinarily he allows from two to twelve hours' delay in the residence for ceremonies, etc., when the body must go to the waiting mortuary, where it remains for seventy-two hours longer, under medical observation, when the mortuary physician gives his certificate, if all goes without unforeseen incidents, and the interment takes place in the adjoining cemetery.

Thus it is seen that there are, with the *Leichenfrau,* four independent expert inspectors. All are on the *qui vive* in carrying out the system, which is popular and understood by all classes. The waiting mortuary consists of a main hall, where the bodies lie in open coffins, embowered by plants in the midst of light, warmth, and ventilation. There is also a laboratory equipped with apparatus for resuscitation, *post-mortem* room, separate rooms for infectious cases and accidents, a chapel, and quarters for the physicians and attendants.

Count Michael von Karnice Karnicki, formerly chamberlain to the Czar of Russia, invented, in 1898, an exceedingly clever apparatus for the prevention of premature burial. Being firmly convinced that thousands of persons are buried every year while in a state of lethargy, he prepared a system of signalling, which has been adopted in one or two instances, but only, so far as I know, in Europe.

In this invention, a tube protrudes about four feet above the surface of the grave, and, upon the top of it, is fixed a small metal box with a spring lid. To the lower end of the tube, which just enters the upper lid of the coffin, is fixed an india-rubber ball, charged so fully with air that the slightest pressure upon it would result in the discharge of this air through the tube. This would release the lid of the box, which is adjusted to spring open at the slightest pressure. Moreover, the opening of this lid would automatically raise a small flag, and, at the same time, would start an electric bell, not only over the grave, but in the sexton's house as well. Under this system, the slightest suggestion of breathing on the part of the supposedly dead person, or the smallest movement of the body, would suffice to open the box, raise the flag, and sound an alarm, while the additional mechanism in the tube would immediately begin to pump air down to the interior of the coffin, that the person who has been buried by mistake might be preserved from suffocation until such time as assistance might arrive.

On March 1, 1909, the House of Commons ordered to be printed for distribution "A Bill to Amend the Law Relating to the Registration of Deaths and Burials." The Committee of Examination confessed themselves "much impressed" by the weight of the evidence brought before them, tending to show that the current medical examinations were insufficient; and they ordered a more thorough and complete examination and certification in the future.

CHAPTER VII

BURIAL, CREMATION, MUMMIFICATION, ETC.

ALTHOUGH burial is an extremely unhygienic and unwholesome custom, it is a practice that is common to all Christian countries. Originating in the popular faith in the doctrine that the body should be preserved that it might arise in its entirety at the "day of judgment," this idea, though now seldom advanced as an excuse, is at the bottom of the antipathy that is so frequently shown in regard to cremation. Of course, it is needless to say that, as the process of putrefaction soon returns the physical body to the dust of the earth, through which it passes again into all forms of vegetable and animal life, the impossibility of any sort of bodily resurrection, without the performance of a more stupendous miracle than the human mind could imagine, is obvious.

In fact, the only argument that can be advanced in favour of the practice of burying the dead, as against that of cremation, is based upon the principle that a buried body may be exhumed—after a considerable space of time has elapsed, if necessary—and the effects of poisons, etc., traced—murderers frequently having been brought to justice by this means when they would have undoubtedly escaped punishment for their crimes, if the most convincing evidence against them could have been destroyed by fire. Such an argument as this, however, weighs but little against the many great advantages that would be derived from the practice of cremating bodies, for cremation is so manifestly the only wholesome and hygienic method of disposing of the dead that it should be legally adopted by all nations calling themselves civilized.

Incineration, or cremation, was the ancient Roman method of reducing the body to ashes, but the ancient Jews early adopted the custom of burial. Thus Abraham, in his treaty for the cave of Machpelah, expressed the desire to secure a suitable place in which "to bury his dead out of his sight"; and about the only records of burning the dead that we find in the history of the Jewish people are (1) the case of Saul and his sons, whose bodies were undoubtedly too badly mangled to be given the royal honours of embalmment, and (2) the burning of those who died of the plague, a sanitary measure apparently adopted to prevent further spread of the contagion.

As all nations of the ancient civilization held that it was not only an act of humanity but a sacred duty to pay great honours to the departed, the burial and funeral rites were frequently of a very elaborate character. Among the Hebrews, these began with the solemn ceremony of the last kiss, and, after the eyes had been closed, the corpse was laid out and perfumed by the nearest relative, and the head, covered with a napkin, was subjected to complete ablution in warm water, a precaution that was supposed to make premature burial impossible.

While the Jews frequently embalmed the body of the dead, in no part of the world was this rite performed so scientifically as in Egypt. When great personages like Jacob and Joseph died, the greatest care was exercised in embalming them, but there is no means of ascertaining whether the earlier generations followed this practice, or simply buried their dead in caves, or in the ground. We know, however, that the elaborate process followed in later years was finally abandoned for a simpler and less effective method—that of merely swathing the corpse round with numerous folds of linen and other stuffs, and anointing it with a mixture of aromatic substances, of which aloes and myrrh formed the principal ingredients. To be sparing in the use of spices on such an occasion was regarded as a most discreditable economy, for the profuse use of very costly perfumes was regarded as the highest tribute of esteem that could be paid to the departed. In view of these facts, it is easy to believe the writer in the *Talmud* who tells us that no less than eighty pounds of spice were used in burying Rabbi Gamaliel, and Josephus reports that, at the funeral of Herod, five hundred servants were in attendance as spice-bearers. From the narrative in the New Testament we see that a similar custom was followed at the burial of Christ.

The Jews, like most Oriental nations, were given to the most inordinate exhibitions of grief. From the moment that the vital spark was known to have departed from the body, the members of the family, especially the women of the household, burst forth into the most doleful lamentations, upon which they were joined by neighbours and relatives, all of whom crowded to the house as soon as they heard of the bereavement. By the more aristocratic anything like outside participation in the grief of the family was forbidden, and, instead, this service was performed by certain women who were known as public or professional mourners. When engaged, they seated themselves in the family circle, and, by studied dramatic effect and eulogistic dirges, excited greater lamentations on the part of the immediate family. Sometimes instrumental music was also introduced.

As in all Oriental countries, burial among the Jews occurred more quickly after death than is generally the practice in this country. Even when the body had been carefully embalmed, interment was not long delayed, and, when this precaution had not been taken, it was invariably held within less than twenty-four hours. This was partly due to the climatic conditions, and partly to the circumstance that the Jews taught that anybody who came near the death chamber was unclean for a week.

The casket, or coffin, is the invention of the Egyptians, but the Jews and some other races early adopted it. Originally these chests of the dead were composed of many layers of pasteboard glued tightly together; later they were of wood, or stone, but for the most eminent men was reserved the honour of being buried in coffins of sycamore wood.

Although the bodies of the dead were sometimes placed in these caskets before being transferred to the grave, the most common method of transporting the corpse from the home of the family to the place of interment was by means of a bier, or bed, which was sometimes composed of very costly materials. Instances are known in which kings and extremely wealthy personages have been conveyed to their tombs on their own beds, but the bier in common use among the poorer classes was usually little more than plain wooden boards, fastened to two long poles, and on which, concealed by a sheet, or other thin coverlet, the body lay. It is just such a bier as this that is described in the Bible, and they are still used in all Eastern countries.

When the deceased was of humble position, none but the relatives and close friends attended the obsequies, unless the family could afford to employ the public mourners and their minstrels, in which case the latter walked before and around the bier, frequently lifting the coverlet and exposing the corpse, which was always a signal to the company to renew their shrill cries and doleful lamentations. Thus, at the magnificent funeral of Jacob, these mercenaries maintained an almost ceaseless expression of the most passionate grief, and when the boundary of Canaan appeared—the site of the sepulchre—the entire company halted, and, for seven days and nights indulged in these violent exercises of mourning under the leadership of the host of professionals who had been employed for the occasion.

Although sepulchres have long been in use in Eastern countries, even the ancient races seldom made the mistake of erecting them in close proximity to human dwellings. No matter how elaborate they may have been—and from those that are still left it is easy to imagine that money was not spared when some of the tombs were constructed—the health regulations of the time required that they should be built without the precincts of the city. Among the Jews—as shown in the regulations of the Levitical cities—it was specified that the distance should not be less than 2000 cubits from the city walls. Jerusalem alone was excepted, and even there, this privilege was reserved for the members of the royal family of David, and some few others of exalted distinction.

During the first three centuries of the Christian era this custom remained unchanged. The Emperor Theodosius issued an edict expressly forbidding the burial of the dead within any town, whether in churches or not, and Chrysostom not only confirmed this view, but when the Donatists buried their martyrs in churches, they were obliged to remove them. Even in the fourth century, when the building of oratories, or chapels, over the remains of eminent Christians—martyrs, prophets, etc., —began, the canon law held this practice to be unlawful, and it was not until the sixth century that this statute was to any great degree disregarded.

Thus, it will be seen that while the Roman nation continued to maintain the custom of cremation, the Christians adopted the practices of the Jews, and buried their dead. St. Augustine, in several passages, commends this custom, not for the reason—he says—that we are to infer that there is any sense or feeling in the corpse itself, but simply because we are to believe that even the bodies of the dead are under the providence of God, to whom such pious offices are pleasing, through faith in the Resurrection.

Cremation

The idea of having hundreds and thousands of decaying bodies in the immediate vicinity of human habitations should be so repellent to any sensible person that argument ought to be unnecessary. The only point that can be urged against this practice, as I have already said, is that, in certain cases, it is important, from a medico-legal point of view, to have the body where it can be exhumed, if necessary.

It seems important that we should insist as strongly as possible upon this point— the logical necessity of cremating the dead. Rightly considered, this practice does not in any way conflict with Christian teaching, but conforms to its highest standards. After death, we are not concerned with the material man, but with the spiritual replica (granting anything to exist at all), and no one in these enlightened days would think for a moment that a truly physical resurrection of the body took place. It would not be desirable, in the first place, and is an obvious impossibility, in the second. Yet it is only this worn-out and effete tradition of physical resurrection which prevents the general adoption of cremation—the far more sanitary and rational process. Let us consider some of its benefits a little more closely.

In the first place, then, there is the very evident reason that there will not be enough space, very soon, to contain all the bodies that are to be consigned to the earth. The population of Brooklyn and of New York (to take typical cases) increased more than seven times in fifty years—from 1840 to 1890, and the population is now more than five million. And, as Mr. Augustus G. Cobb well says: [1] —

"The effect, in twenty years, on these six cemeteries will be to increase by a million additional bodies the 1,336,000 already received. Brooklyn is twenty-three times as large today as it was fifty years ago, when the first interment was made in Greenwood; and, as a natural consequence, this cemetery, once surburban, has become intramural. It need surprise no one to learn that its exhalations have been complained of in South Brooklyn, and, considering the thousands annually interred within its grounds, and the increasing density of population, we can readily believe that the evil, instead of diminishing, will increase. . . . "

It is needless to point out that such a mass of decomposing organic material, so close to the very homes of

[1] *Earth Burial and Cremation*, pp. 26, 27.

the inhabitants, is apt to prove extremely dangerous; first, by contaminating the wells, springs, and water in the neighbourhood; and secondly, by vitiating the atmosphere and rendering a serious epidemic not only possible but exceedingly likely. When we know that germs can be carried through the air for miles—as they can—the immediate peril of a graveyard need hardly be pointed out. As Sir Lyon Playfair (after making a most exhaustive investigation of the whole question) expressed it:—

"In most of our churchyards the dead are harming the living by destroying the soil, fouling the air, contaminating water-springs, and spreading the seeds of disease."

Says Mr. Cobb:—

"Opposition to incineration springs chiefly from ignorance of the manner in which it is effected; and to remove all misapprehension, it cannot be too distinctly stated that the body *never* rests in flames, while during the entire process there is no fire, no smoke, or noise to grieve in any manner the bereaved. The consuming chamber in which the body is placed is built of fireclay, and is capable of resisting the highest temperature. Under it and around it the fire circulates but cannot enter in. The interior, smooth, almost polished, and white from the surrounding heat, presents an aspect of absolute, dazzling purity; and as the body is the only solid matter introduced, the product is simply the ashes of that body. During the entire process of incineration the body is hidden from view. . . . The heated air soon changes it to a translucent white, and from this it crumbles into ashes."

Is not this picture far more pleasant than that of the grave? Is it not far more cleanly, hygienic, and sensible? Is it not obvious that cremation is simply unpopular for the reason that it is based on a mass of false sentiment and worn-out theological dogmas as to the resurrection of the body? From every rational point of view, everything is in favour of the process, nothing against it.

I am familiar with the so-called objections to cremation advanced by M. Jean Finot in his *Philosophy of Long Life*. I can only say that they appear to me, for the most part, totally inadequate. Some of his *facts*, it is true, are worthy of serious consideration: his negative evidence as to the pollution of the air in the neighbourhood of cemeteries, etc. But his idea that the life of the body is perpetuated in the lives of the worms that devour it!—*that* appears to me little short of absurd. In direct opposition to this view let me quote the opinion of Mrs. Annie Besant, who, in her *Death, and After?* says:—

"One of the great advantages of cremation, apart from all sanitary considerations, lies in the swift restoration to Mother Nature of the physical elements comprising the dense and ethereal corpses brought about by the burning, and hence the quicker freedom of the soul from the body. On the assumption that a soul of some sort exists, this would certainly seem far the more rational supposition; and if materialism be true, and no soul persists, then cremation has the field entirely, since there would remain no valid objection to the practice whatever."

EMBALMING

Embalming is a method of preserving bodies by injections and dressings, both internally or externally applied.

This term is generally given to the process employed by the ancient Egyptians and others, by which corpses were preserved as mummies. The practice is very ancient, and is probably founded on religious rites and observances. The Egyptians believed that it would be possible for the departed spirit, at some future time, to reanimate the body of the deceased, and hence took great pains to preserve it as perfectly as possible. Some of the processes employed were very elaborate and expensive, and could only be afforded by the wealthy. The most elaborate process was somewhat as follows:—

A deep cut was made beneath the ribs on the left side, and through the opening thus made the internal organs were removed, with the exception of the heart and kidneys. The brain was also extracted through the nose by means of a bent iron instrument. The cavities of the skull and trunk were washed out with palm-wine, and filled with raisins, cassia, and similar substances; and the skull was dressed by injecting drugs of various kinds through the nostrils. The body was then soaked in natron for seventy days. It was then removed and wrapped carefully in linen cloth, cemented by gums.

The less expensive process consisted in removing only the brains and injecting the viscera with cedar oil. When the body was soaked in natron for the same period of time (seventy days), the viscera and soft parts came away *en masse,* and only the skin and bones were left.

The very poor, who could not afford either of the above methods, embalmed their dead by washing the body in myrrh and salting it for seventy days. The body, thus embalmed, was ready for the sepulchre; but it was often kept at home for a considerable time afterwards, and was produced on certain occasions—such as a dinner-party!—and carried round the room "to remind the diners that death was ever with them."

Doubtless the method of embalming differed greatly in different countries and in the same country at various times. The above process was described by Herodotus in his writings as being practised in Egypt at that time. Animals were also embalmed, especially those held to be sacred. It is certain, however, that only a small percentage of the dead organic matter could have been disposed of in this manner; and it is not known what became of the remainder or what disposition was made of it.

Embalming is carried on at the present day, but for very different reasons and in a different manner. The object is not to preserve the body for centuries, as the Egyptians hoped to do, and in fact actually succeeded in doing. In some countries the use of salts of arsenic, corrosive sublimate, etc., is prohibited by law for medico-legal reasons; but embalming can only be performed by toxic substances. Many of these have been tried, with limited success. Essential oils, alcohol, cinnabar, camphor, saltpetre, pitch, resin, gypsum, tan, salt, asphalt, Peruvian bark, cinnamon, corrosive sublimate, chloride of zinc, sulphate of zinc, acetate of aluminum, sulphate of aluminum, creosote, carbolic acid, etc., have all been recommended by modern embalmers. In these days details of procedure vary, but all must conform to the law.

The length of time which a body will keep before decomposition sets in varies greatly. In those cases in which but little flesh is left on the bones, and when the blood has decreased greatly in volume (for example, in consumption, where great emaciation has taken place before death), the body will keep far longer than one which has a large amount of tissue still upon the bones and a large volume of blood. Blood being the active principle in decomposition, its prompt removal is necessary in cases of this character. The time of year, the disease from which the person died, etc., all have an appreciable effect upon the length of time the body will naturally take to decompose; and hence all these factors must be taken into account by the embalmer in selecting the amount and the strength of the fluid to be injected into the arteries of the corpse.

The general procedure is somewhat as follows:—The body being laid out, an incision is made with a sharp knife, and the artery is drawn to the surface by means of a metal hook. The artery selected varies, some embalmers choosing the brachial artery, others the axillary artery, etc. It depends upon the individual choice of the embalmer. If a visible scar is objected to, the brachial artery cannot be used. After the artery is brought to the surface and cut, the embalming fluid is forced into it by means of a small pump provided with two valves, after the manner of the heart. It is intended, indeed, to take the place of the heart in forcing the blood through the body. One of these valves forces the fluid into the artery; the other sucks up the fluid from the bottle in which it is contained. The fluid passes directly across to the heart and other vital organs, and when this has been done a second incision is made just below the heart, which is punctured. The blood is then drawn off from the heart, and the double process is continued until all the blood in the body has been replaced by the embalmer's fluid. Sometimes a second artery is cut in the leg. If the fluid is found to come away clear at this point, without an admixture of blood, the body is clear of blood—the chief decomposing agent.

The fluid which is injected into the body has a tendency to harden the tissues, and they could be made actually brittle if enough were used. The embalmer uses his judgment as to the strength of the fluid. Generally, an 8-ounce bottle of prepared embalming fluid is mixed with half a gallon of water, this being the typical "embalmer's solution."

From a medico-legal point of view, there is much that can be said against embalming. Brouardel has pointed out that embalming can only be performed with toxic substances, and this fact would vitiate any subsequent investigations that might have to be made—in a poison case, for example. Embalming might preserve bodies a much greater length of time than would otherwise be the case; but what is the object to be gained thereby? The body must ultimately decompose, whether embalmed or not, and of what use is the preservation of bodies? Our chief object should be, *not* to preserve them, but to get them out of the way as speedily and as hygienically as possible. It is surely a more pleasing thought to think of a cremated body than to dwell upon the condition of one that has been buried six months or a year.

Mummification

The mummified bodies of some of the Egyptians have doubtless been seen by every one. So perfectly have some of these bodies been preserved that even the features can be recognized after more than three thousand years. The bandages wrapped round the bodies were doubtless antiseptic in character; but the details of their methods have been lost.

Apart from such cases, mummification of bodies may

sometimes take place spontaneously, and the body be mummified instead of decomposing. This is especially the case in dry, hot countries, where there is but little moisture in the air. In the sandy soil of Mauritius, e. g., it is asserted that bodies frequently become mummified. Where there is a lack of air, the body will also occasionally assume this condition, even in our climate. M. Audouard reports a case of a mummified body, discovered by him, in which "the skin was like parchment, shrivelled, and of a buff color. When it was tapped with the back of a knife, it resounded like cardboard." The body had become very light. M. Audouard found also that the skin was perforated with a number of holes, like a colander, and that dust from within escaped through these little holes. A thigh of the leg weighed just one-third of the normal weight. The body had been devoured by mites, which had eaten all the tissues of the woman. The dust within the hollow and mummified limb consisted mostly of the excretions of the mites.

It is asserted that mummification of the body of an unborn child will take place, if the child be preserved *in utero,* and no air is allowed to enter the uterus. It is sometimes seen in the young, more rarely in adults.

Lately, when visiting the Egyptian room in the British Museum, I noticed very carefully the physical peculiarities of some of the excellent specimens of mummification there exhibited. One case was especially interesting. A hand and arm, stripped of the winding bandages, is shown—perfect in its texture, all the nails, and even the texture of the skin, being clearly visible. The arm is shrunk to about one-fourth its normal size (it is merely the skin stretched over the two bones of the fore-arm). The hand is partly clenched. The whole is jet black, and has the appearance of being made of unpolished ebony. The human, living arm has now become petrified, as it were, and takes on the exact appearance of wood. The arm is extremely hard and brittle—so much so, indeed, that it is *cracked* along its upper surface—just as a piece of wood might be cracked or split. This struck me at the time as a very remarkable phenomenon—apparently showing the ultimate tendency of such organic substances to petrify, become coal-like and finally return to the mineral elements from which they sprang.

M. Mégnin divides the work of the "labourers of death" into four periods. In the first, quaternary compounds are attacked and destroyed; in the second, fatty substances are attacked; in the third, the soft parts are liquefied; lastly, in the fourth period, the dried-up mummy is filled with mites.

In all cases (with the exception of cremation), a fermentation takes place before the body is completely destroyed; gas is produced, and the organism is returned to the mineral kingdom more or less rapidly—the rapidity and character of this return being governed by several considerations. This is the invariable process, except in those rare cases in which the body remains frozen, or where it is devoured by wild animals or birds of prey. When the body is immersed in the ocean, it is soon devoured by sharks, crabs, and other carnivorous sea-creatures.

The body, when lying in peaty soils, or when surrounded with other antiseptic influences, will mummify. The body must be rather thin and juiceless, however. There is a church at Toulouse where the structure of the place seems to cause mummification of bodies, owing to a current of air being always present.

The process of preserving the body by *drying,* which has sometimes prevailed among savage people, is probably somewhat similar to the method of preserving meats which is practised by the natives of certain parts of South America. As described by Charles J. Post, the artist and explorer, this is as follows:—

"The national food of the country is the 'chalona' and the 'chuno.' These are consumed so generally that there are many villages east of the Andes in which the people have no other means of support than that which is afforded by the preparation of these edibles. The 'chalona' is nothing more or less than mutton that has been dried so thoroughly that it bears a close resemblance to a mummy. The natives take the carcase of the sheep up into the mountains—sometimes 2000 feet or more above sea level—and there they let it lie all day beneath the rays of the sun. When the dew begins to fall, or there is any apparent dampness in the atmosphere, they cover it securely, and do not expose it to the air again until these conditions have disappeared. When fully preserved under proper conditions, the carcase of the sheep will not weigh more than ten or twelve pounds.

"The Indians eat this meat raw, masticating it to a degree that corresponds to our modern method of 'Fletcherism.' If it is to be cooked, however, it is necessary to stew it for fully ten hours the day before it is to be used, and to boil it again for not less than four hours the day that it is to be served. The natives eat it in combination with the 'chuno'—potatoes that have been treated in the same fashion until they have been dried to about the size of a bantam egg."

As Mr. Post suggests, this is practically a process of mummification.

CHAPTER VIII

OLD AGE: ITS SCIENTIFIC STUDY

THIS subject is of great interest as possibly throwing some light on the question of natural death. Certainly it is a question that should receive the closest attention from scientists. M. Metchnikoff, of Paris, has given it much thought, and I shall have occasion to mention his work immediately. First, however, a few preliminary remarks.

There can be no doubt that the average length of life of the human race should be far greater than it is now. Every animal is supposed to live at least five times as long as it takes to mature; this is the all but invariable rule in the animal world, and should hold good for man also. He matures about twenty, let us say. According to our rule, therefore, he should live to be a hundred, and that without growing decrepit or without being regarded as exceptionally old or long-lived! That *should* be his normal age limit. But, instead of this, what do we find? That the average duration of human life is a fraction over forty-five years; and, more than that, these forty-five years are filled with grievous diseases and illnesses of all sorts, instead of health and happiness. Something is assuredly wrong somewhere; life is far shorter in duration than it should be, and practically every one dies prematurely. The great majority die either from some disease or from some "sudden death," which, as I have shown, is not really sudden death at all, but the sudden culmination of an unobserved diseased condition. Of course, all such persons do not die *naturally*, and it is probable that very few indeed do die what might be called a "natural death." Physiology has even questioned why the body should *ever* wear out, provided the organs remain sound and health be maintained! Thus Hammond stated that "there is no physiological reason at the present day why men should die." G. H. Lewes, in his *Physiology of Common Life*, also said: "If the repair were always identical with the waste, never varying in the slightest degree, life would then only be terminated by some accident, never by old age." Dr. Munro asserted that "the human body as a machine is perfect . . . it is apparently intended to go on for ever." Dr. Gregory, in his *Medical Conspectus*, wrote: "Such a machine as the human frame, unless accidentally depraved or injured by some external cause would seemed formed for perpetuity." Other authors could be quoted to like effect. Mr. Harry Gaze, indeed, devoted a whole book to this question, and tried to show why we need *never* die if we only make up our minds to stay alive![1] The arguments against this position have been given elsewhere.[2]

At all events, it is certain that the great majority of persons do die prematurely. The greatest number of such premature deaths are from diseases of various kinds. Such causes of death are analysed and classified in a little book entitled *Premature Death*. Here we read that nine-tenths at least of all deaths are premature! (p. 5), and this is doubtless short of the truth. All accidental deaths are, of course, also premature; so

1 *How to Live Forever.*
2 *Vitality, Fasting and Nutrition*, pp. 328–29.

that the margin of cases of natural death is small indeed. It is amazing, when we consider this fact, that so little attention is paid to it either by doctors or the public. However, this is not the place to consider that question.

On page 14 of the book just quoted, the author makes the following assertion:—

"With the completion of manhood, diseases indicative of local degenerations of tissue begin to be predominant, and, with each successive stage of life, this predominance becomes more marked. In old age the degenerative changes, which at earlier periods of life are regarded as the signs of disease, now appear as the natural consequences of decay, and death becomes a *physiological,* not a pathological fact—as the termination of a natural life, not as the premature close of a life cut short by disease."

Metchnikoff however, takes the opposite view very strongly. He says, in part:—"It has often been said that old age is a kind of disease. . . . In fact the great resemblance between these states is incontestable. . . . The theory of old age and the hypotheses which are connected with it may be summarized in a few words: The senile degeneration of our organism is entirely similar to the lesions induced by certain maladies of a microbic origin. Old age, then, is an infectious chronic disease which is manifested by a degeneration, or an enfeebling of the nobler elements, and by an excessive activity of the macrophages."[1]

Metchnikoff holds that, if death were due to old age, it would be sought for and anxiously awaited (instead of being dreaded and feared), just as we long for a night's sleep after a day of hard and trying work. It is prob-

[1] ***Old Age,*** by Elie Metchnikoff. ***Smithsonian Report,*** pp. 542–48.

able that this is the case. It is probable that nature intended just such a plan. The dread of death that is so universal merely shows us that, in practically all cases, death has been *premature;* it has come before it was wanted—before its appointed time. There is every reason to believe, and every analogy points to the fact, that death should be welcomed, as sleep is welcomed, by those fatigued. Metchnikoff adduced some cases in support of his contention; and he is probably right in his central claim.

Old age is invariably regarded as a period of decrepitude and mental imbecility. And although this is, as a matter of fact, the all but invariable rule, there is no real reason why such should be the case. Hardly any of the wild animals show signs of decrepitude in a similar manner, and only some of the domestic animals do. The rule would seem to be that the closer we live to nature, the longer is death postponed, and the more painless and sudden it is. Those living as the majority do, and indulging to an unlimited extent in rich foods, dissipations of all sorts, and what are generally known as the "good things of this world," *do* degenerate prematurely and lose their mental and moral fibre, no less than their physical bodies. Decay is the rule; uselessness is the general condition of the aged with most civilized nations—and even of some that are not civilized! The inhabitants of Tierra del Fuego, for example, kill their old women before they kill their dogs, when they are threatened with famine. When asked why they do this, they reply: "Dogs catch seals, while old women do not!" Although civilized nations do not adhere to the doctrine of survival of the fittest so relentlessly, they nevertheless show by word and action very frequently that they wish the day would come when such "nuisances" will be removed.[1]

I may now give a brief survey of what is known of old age and its causes, and some of the theories that have been advanced from time to time to explain its phenomena. Very few of these need be considered, as they are not either clear enough or comprehensive enough to deserve such discussion. A few, however, are very ingenious, and deserve careful consideration.

Certain authors have advanced what might be called a "psychological" theory of old age and death. One grows old and dies when there is no longer an incentive to live. As Dr. E. Teichmann expressed it: "They grow old because they are no longer occupied with life."[2] This theory would completely fail to account for the phenomena of old age, even if it succeeded in accounting for death. There are many pathological, degenerative phenomena connected with old age which must be taken into account in this connection—degenerations which are not apparently due to any psychic causes, but to purely physical conditions. Such a theory would be no means explain the facts.

Says Metchnikoff:—

"Numerous scientists affirm that old age finally results because it is impossible for an organism to repair the cellular losses by the formation of a sufficient number of new elements—that is to say, because of the exhaustion of the reproductive faculty.

"One of the scientists who has more especially concerned himself with general questions, Weismann, expresses himself

1 For the practical side of this question—the reader is referred to my little book *Death Deferred:* How to Live Long and Happily, Defer Death and Lose all Fear of It.

2 *Life and Death*, p. 145.

on this subject in a very categorical manner. According to him, the senile degeneration that ends in death does not depend on the wearing away of the cells of our organisms, but rather upon the fact that cellular proliferation, being limited, becomes insufficient to repair that loss. As old age appears in different species and different individuals at various ages, Weismann concludes that the number of generations that a cell is capable of producing differs in different cases. It is, however, impossible for him to explain why, in one example, cellular multiplication may stop at a certain figure, while in another it may go much further.

"The theory appears so plausible that no attempt has been made to support it by precise facts. We even see, in the most recent attempt at a theory of old age by Dr. Bühler, the thesis of the exhaustion of the reproductive power of the cells accepted and developed without sufficient discussion. It cannot be denied that it is during embryonic life that cells are produced with the greatest activity. Later on this proliferation becomes slower, but it nevertheless continues throughout the course of life. Bühler attributes the difficulty with which wounds heal in the aged precisely to the insufficiency of cellular production. He also thinks that the reproduction of the cells of the epidermis, which are to replace the desiccated parts of the skin, diminishes notably during old age. According to this author, it is theoretically easy to predict the moment when cellular multiplication of the epidermis must completely cease as the desiccation and desquamation of the superficial parts continue without cessation, it becomes evident that it must finally result in the total disappearance of the epidermis. The same rule is applicable, according to Bühler, to the genital glands and muscles, and all sorts of other organs."—*Old Age,* pp. 538–39.

Metchnikoff advances several arguments against this theory—none of which, to me, appear altogether conclusive. A much stronger argument against this original-stock-of-energy theory is to be found in such a case as the following: A person has an attack of sickness, and almost dies. He comes as near as it is possible to dying, without actually doing so. Recovering, however, he lives on for half-a-century, in comparatively good health. Now, at the time of the illness, if that person had died, the reproductive power of his cells would have been lost for ever; and yet, simply by reason of the fact that he turned the critical point and recovered, his cells continue to reproduce for half-a-century longer! Surely, we must give up the notion that the potential energy of the cell is inherent at birth in such a case, and assume that some *new* stock of energy is imbibed from some external source, sometime later on in life? The idea that the diseased cell, all but dead, possessed the potential energy to reproduce for fifty years, while still in that condition, seems too absurd to need criticism.

Bichat says that:—

"In the death which is the effect of old age the whole of the functions cease, because they have been successively extinguished. The vital powers abandon each organ by degrees; digestion languishes, the secretions and absorptions are finished; the capillary circulation becomes embarrassed; lastly, the general circulation is suppressed. The heart is the *ultimum moriens*. Such, then, is the great difference which distinguishes the death of the old man from that which is the effect of a blow. In the one, the powers of life begin to be exhausted in all the parts, and cease at the heart; the body dies from the circumference, towards the centre; in the other, life becomes extinct at the heart, and afterwards in the parts; the phenomena of death are seen extending themselves from the centre to the circumference."—*Recherches physiologiques sur la Vie et la Mort* (p. 143).

These conclusions were confirmed by a number of cases cited by Dr. John D. Malcolm in his *Physiology of Death from Traumatic Fever.*

Other writers have attacked this problem in a different manner. They, too, have contended that old age and death are due, in a sense, to the decrease of the vitality of the body, but have asked themselves the question: Why should this vitality become lessened with old age, seeing that it is (supposedly) constantly being replaced by a fresh stock of vitality from the food which is continually eaten? On the theory commonly held, the bodily energy is supposed to come from the food we eat, and that is constantly being supplied to the system—in old age, as in youth. Why, then, should these degenerative changes take place, and the vitality decrease? These authors have come to the conclusion that the vitality depends upon the *state* of the body—its *health;* and, so soon as the body becomes clogged and blocked; as the result of wrong food-habits and other causes, old age, premature decay, and death result.

Two writers who have taken this view are Dr. Homer Bostwick, who published his *Inquiry into the Cause of Natural Death; or, Death from Od Age,* in 1851; and Dr. De Lacy Evans, M.R.C.S.E., who issued his book, *How to Prolong Life; An Inquiry into the Cause of Old Age and Natural Death,* about 1880. The similarity of the views of these two authors is very remarkable, but each apparently wrote in ignorance of the work of the other—one in America, the other in England. Yet their views are almost identical. Both authors contend that "induration and ossification are the causes of old age and natural death." Both contend that these are the true *causes,* and not the *result,* of old age. Both these authors contend, further, that this induration and ossification are due to the excess of lime and other earthy salts that have accumulated within the system as life progressed; that old age advances just in proportion to the amount of this earthy matter in the system, and that old age is retarded just to the extent that it is kept out. But since all such substances can only be introduced into the system through the food and drink, they sought to find these foods which contained the minimum of such earthy matter, and these they found to be *fruits.* By living on fruits, then, they were enabled to retard the progress of old age and natural death, both in themselves and in all others who undertook to follow their diet. Careful analysis of the various foods confirmed their theory, which was also supported by a number of experimental facts. They therefore concluded that this was man's natural diet—that best suited to his body; and that, by eating fruit, man could very largely retard the oncoming of old age and natural death.[1]

These authors made the degree of the vitality depend upon the condition of the body—and hence upon the food. On the other hand, it must be remembered that the utilization of the food, and its successful elimination, will depend upon the degree of vitality present—i. e. the vitality will depend upon the state of the body, and the state of the body will depend upon the degree of vitality. We are here, therefore, in a vicious circle. Nevertheless, I think that these authors have attacked the problem in the right way, and I shall have occasion to recur to their views later on, when I come to consider this question of the relation of health to vitality.

[1] I have defended the fruitarian diet at considerable length in my book *The Natural Food of Man.*

There are also many facts which support such a view. Let us consider some of these.

The most marked feature in old age is that a fibrinous, gelatinous, and earthy deposit has taken place in the system—the latter being chiefly composed of phosphate and carbonate of lime, with small quantities of sulphate of lime, magnesia, and traces of other earths. The accumulation of these solids in the system is doubtless one of the chief causes of ossification, premature old age, and natural death. In the *bones* this is most noticeable. the amount of animal matter in the bones decreases with age, while the amount of mineral matter increases. This is especially marked in the long bones and the bones of the head. They thus clearly show us that a gradual process of ossification is going on throughout life.

As age advances the *muscles* diminish in bulk, the fibres become rigid and less contractile, becoming paler and even yellowish in colour, and are not influenced by stimuli to the same extent as in youth. Tendons also become ossified to a certain extent, while there is a diminution of the fluid in the sheaths of the tendons. The brain increases in size, up to about forty years of age, when it reaches its maximum weight. After this period there is a gradual and slow diminution in weight of about one ounce in every ten years. According to Gazanvieilh, "the longitudinal diameter of the brain of an old man, compared with that of a young man, is six inches one line, French measure, for the former, and six inches four lines for the latter; whilst the transverse diameter is four inches ten lines in the old man, and five inches in the young man." The convolutions of the brain become less distinct and prominent.

The *dura mater* is often found apparently collapsed or

corrugated. It is thickened and indurated, and ossific deposits on the arachnoid surface are very common. The membrane is sometimes found to have an abnormal dryness; the arteries supplying the brain have, in old age, become thickened and lessened in calibre; the supply of blood thus becomes less and less, leading to the mental imbecility of the very aged. This gradual process of degeneration in the arteries, not only in the brain but throughout the body, is well recognized, and is perhaps one of the most important of all the changes that take place in old age. So important a symptom is it considered, that it has given rise to the old saying that "a man is as old as his arteries." The capillaries also become choked or blocked and clogged up, as the result of the earthy matter accumulated in the system.

These changes taking place in the arteries, greater pressure is thrown upon the *veins*, which dilate, their coats becoming thinner, and they even become tortuous and varicose.

The gradual process of induration and hardening going on throughout the system is noticeable also in the *heart*—giving rise to various affections known to us under a variety of symptoms. The *lungs* gradually lose their elasticity, and increase in density. The *air-cells* and *bronchi* become dilated—hence emphysema and chronic bronchitis are so often seen in the aged.

The *salivary* glands become hardened, and decrease in bulk. The saliva is either secreted in large quantities, so that "dribbling" takes place, or in quantities so small that the mouth is hardly moistened. These changes are probably due in part also to lack of central inhibition.

In the *stomach* the gastric juice is secreted in a diluted form, and is deficient in pepsin; moreover, the muscular walls of the stomach gradually lose their wonted contractibility; the peristaltic motion becomes weak; chyme is imperfectly manufactured, and all the processes of digestion weakly performed.

The *liver* shows the effects of old age by its imperfect bile-forming qualities. Fatty matters are not thoroughly emulsified or absorbed by the lacteals—though this may be due to an alteration in the fluid secretion in the pancreas.

In the *intestines,* the small vessels which supply the follicles and various glands become indurated, or even clogged up, in old age. The walls of the intestines become opaque, and lose their contractibility, while the *villi* containing the lacteals undergo the same gradual alteration. It will be seen from the above how necessary it is that all food should be restricted in quantity and simplified in quality in old age! Almost all the *viscera,* and particularly those glands and organs connected with the sexual apparatus, show signs of old age. The walls and structures become harder in texture, and less pliable.

In the *eye,* in old age, there is diminished secretion of the aqueous fluid in the anterior chamber, the cornea becomes less prominent, the pupil becomes more dilated, from lessened nervous sensibility—hence distant sight and the indistinct and confused view of near objects in the aged. Cooper states that the retina, in old age, is found "thickened, opaque, spotted, buff-coloured, tough, and in some cases ossified." Quain calls attention to the fact that the colour, density, and transparency of the lens presented marked differences in different periods of life. In old age it becomes flattened on both surfaces, and assumes a yellowish or amber tinge. It loses its transparency, and gradually increases in toughness and in specific gravity. Cataract is rarely found in the young, but frequently in the aged.

The *ear* is subject to the same gradual process of ossification. The cartilages of the external ear become hardened, or even ossified; the glands which secrete the ear-wax undergo the same alterations as are found in other glands. The secretion becomes less, and altered in quality. The *membrana tympani* becomes thickened and indurated; the ligaments connecting the ossicles (maleus, incus, and sapes) become hardened, their pliability is lessened; thus vibrations which are already imperfect, owing to induration of the *membrana tympani,* are improperly converted by the ossicles across the cavity of the tympanum, by means of the internal ear (the structures and fluids of which have undergone the same process of consolidation), to the auditory nerve, the sensibility of which decreases with the senile changes of the brain. Hence the impaired and confused hearing so often observed in aged persons.

The whole membrane covering the *tongue* becomes thickened and hardened in old age; its surface becomes dry and furrowed, while the blood-vessels supplying the papillae are decreased in size; hence the sense of taste is diminished.

In old age, the sense of *smell* is lessened, owing to the hardening of the membranes and internal cartilages; moreover, the fibres of the olfactory nerves lose their susceptibility.

The sense of *touch* throughout the body is greatly diminished: this for several reasons. The sensibility of the nerves is lowered, as well as the reactions of the centres. The epidermis becomes thickened and less sensitive. The capillaries supplying the papillae are also lessened in calibre; the action of the various sebaceous glands is also diminished; the skin becomes dry, shrunken, and leather-like. It thus has a cracked and furrowed appearance, and has a tendency to pucker-up. Hence the wrinkles of old persons. In old age the skin contains more earthy salts than in youth.

As is well known, the *teeth* are almost invariably lost before age is far advanced—this being due partly to external causes, partly to the lessening and corruption of the blood supply, upon which the nutrition of the teeth depends. As a result they decay and fall out.

The *hair* is generally lost, and it usually becomes white. The cause of this for a long time puzzled physiologists; Metchnikoff advanced the view that this blanching of the hair is due to the action of certain micro-organisms, which devour the colouring matter. Metchnikoff's theory of the blanching of the hair fails to account for certain facts, however—such as the complete whitening of the hair over night, as the result of purely nervous shock.

The *rate* of the *pulsation* does not vary much as age advances. From eighty to ninety it averages seventy-three to seventy-four in men, and seventy-eight to seventy-nine in women.

The *respiration* of elderly persons is said to average seventeen per minute. This gradually increases and in men, between eighty-five and ninety, it averages nineteen to twenty, and in men over ninety it averages twenty-three. In women it is a little quicker.

One of the most interesting features in advancing age is the *lessening size and weight of the cell multiplying and blood producing organs*—the spleen, the lymphatic glands, and Peyer's glands—coincidently with the lessening of nutritive activity and therefore of the demand upon the blood factors. The *thymus* gland fades soon; its special contribution to the blood compound ceasing possibly to be required a few years after birth. The *lymphatic* glands are large in childhood and youth, and are easily excited to inflame and to enlarge unduly. The *mucous membrane* of the stomach and intestines is usually thin and pale in the aged.

In old age the stock of *vitality* is decreased, but whether this is due to the state of the blood, or of the tissues, or both; or whether the state of the blood and the tissues depends upon the amount of vitality; and whether this vitality can be replenished as life advances, and if so, how; or whether a certain fund of life is inherent in every living organism at birth—which no skill of man can add to—all these are questions which we cannot now discuss. They are treated at considerable length in *Vitality, Fasting and Nutrition,* pp. 225–303, to which I may refer the reader for further details.

The theory has been advanced that we grow old and die for the reason that the brain and nervous system become worn out because of the constant stimuli that have been poured upon them since they began their natural life. They are simply worn out, and refuse to function longer on that account.

There is doubtless a grain of truth in this theory, but it cannot be accepted as in any way an adequate explanation of the facts. For, were it true, it is obvious that those persons who experienced the greatest number of stimuli in their life-times would be the first to wear out, whereas we know as a matter of fact that nearly all persons die at about the same age, no matter how many or how powerful the stimuli to which they had been subjected. Indeed, statistics would seem to show that the busy man of the great city, the mental worker, lives longer than the farmer and the man who lives merely a vegetable existence in the country. Such being the case, it is hard to see how this theory can be made to "hold water."

Then, too, we have the "cometh up as a flower" theory. When we regard the growth, blooming, and death of a summer flower—the shooting upward of the flower stalk of a poppy, for example, with its blossoms, its seeding, and the suddenly ensuing juiceless and dead rigidity, we contemplate phenomena not wholly unlike what takes place in the human organism, when regarded in the large, passing from infancy to maturity and old age.

What has taken place in the poppy stalk?

One class of plant cells has developed, multiplied, and from the products which have issued from them have been produced the stalk proper and leaves. Immediately another class has, in like manner, given rise first to the bud, then to the gorgeous blossom with its stamens and pistils. Fertilization follows in its timed order, and later another class of cells mature as seed.

It has been held that these latter cells in some manner sap and eviscerate, so as to speak, the cells of every other tissue of the plant; and, thus sapping them of their life elements or germs, condense these latter in the seed, where they may long lie dormant, yet capable of producing another plant, and that the parent plant, thus sapped and eviscerated, dies naturally, its life being virtually taken away and carried forward to the seed for another year.

The primary object of all plant life, then, according to this theory, is the perpetuation of the species, and, that object once accomplished, there is no longer any "use" for the plant, which dies at once or soon after. This same idea has been applied to animal life, and even to human beings, and it has been contended that the primary object of living is to bring ever new specimens of the human race into being.

What a hollow mockery! An endless procession of beings with no other aim than to procreate, to perpetuate the species—and to what end? That the offspring may in turn procreate, and thus the farce be kept up for ever! Can we conceive that such is the scheme of nature? Is it not more rational to suppose that the aim and end of living is to create and to enjoy, and that only one function (doubtless an important one, but only one, nevertheless) is to perpetuate the race? Would this not seem to be borne out by the fact that the parents do not die, or apparently even shorten their lives in the slightest degree, by giving birth to children, whereas if this theory were true, that should be one of the cardinal and central features of it? The theory cannot be said to withstand the test of experience or common sense.

As I have said above, most authors are inclined to regard old age as a process of rapid decay and degeneration—e. g. Metchnikoff, quoted before. Some authors, however, are not at all disposed to take this stand. Dr. Charles S. Minot, e. g., is inclined to take an entirely different view of the matter. So far from regarding old age as some sort of disease that is to be avoided, he contends that we age far more *slowly* in old age than we do in youth, and that the rate of decay is in precisely inverse ratio to that generally held to be true. He produces a great mass of evidence in favour of this contention, for which the reader is referred to his excellent and interesting volume (*Age, Growth and Death*). The following quotations may be accepted as exemplifying this author's theory:—

"Rejuvenation is accomplished chiefly by the segmentation of the ovum. . . . As we define senescence as an increase and differentiation of the protoplasm, so we must define rejuvenation as an increase of the nuclear material. . . . If it be true that growing old depends upon the increase of the protoplasm, and the proportional diminution of the nucleus, we can perhaps in the future find some means by which the activity of the nucleus can be increased and the younger system of organization thereby prolonged. . . . We can formulate the following laws of cytomorphosis:—

"First, cytomorphosis begins with an undifferentiated cell.

"Second, cytomorphosis is always in one direction, through progressive differentiation and degeneration towards the death of the cells.

"Third, cytomorphosis varies in degree characteristically for each tissue. . . .

"Finally, if my arguments before be correct, we may say that we have established the following four laws of Age:—

"First, rejuvenation depends on the increase of the nuclei.

"Second, senescence depends on the increase of the protoplasm and on the differentiation of the cells.

"Third, the rate of growth depends on the rate of senescence.

"Fourth, senescence is at its maximum in the very young stages, and the rate of senescence diminishes with age.

"As the corollary from these we have this,—natural death is the consequence of cellular differentiation."

Indeed, as Dr. C. A. Stephens,[1] has said:—

"When we ask the question boldly: Why does the human body grow old, and at length cease from function?—putting

[1] *Natural Salvation*, p. 78.

the inquiry in the bio-physical sense, the answer seems to be that the personal life embodied in the organism is at length overcome and overmatched *by the totality of the resistance to life which it encounters,* from the embryonic stage onwards, more especially by the general telluric resistance, physical, chemical, molar, molecular, which the protoplasmic molecules of the organism meet with as long as they maintain the personal life. After adult age is reached, they lose ground in the struggle, and at last succumb. The downward curve of the somatic cell has begun."

The physiological processes by which food is reduced, comminuted, corrected as to its chemical constituents, peptonized, hepatized, oxygenated, and, in a word, carried forward to higher and higher stages of chemical instability, fit for assimilation by the tissue cells—all these processes set up a heavy draught on the collective animal life of the body, and necessitate the putting forth of energies on the part of all the cells which cause an ever increasing deficit of potential, a growing debt from overwork, a chronic accumulation of the effects of fatigue, which, under present conditions, must sooner or later lead to a running down of the cells.

Under favourable conditions a cell may gain potential; but the severe, steady draught on cellular energy necessary to maintain organic nutrition, even on the best food at present procurable, bankrupts the collective energies of the cells within a century.

In one sense, therefore, it is our food which brings us to death's door—that is to say, the exhausting physiological processes necessary to prepare it for cell nutrition will in the end work the most perfect existent animal organism to death.[1]

[1] See several lengthy discussions of this point in *Vitality, Fasting and Nutrition.*

One of the most ingenious theories of the causation of old age and natural death (and of their possible prevention) is that formulated by Mr. C. A. Stephens, in his book *Living Matter: Its Cycle of Growth and Decline in Animal Organisms.* In this excellent little book the author discusses the various theories of old age, and pretty effectually disposes of them. He then advances one of his own—postulating, at the same time, a possible course of life that would offset physical death—at least, for a very greatly extended period. After showing the improbability of the current notion that we possess a given fund or stock of vitality at birth, which we simply "live out" in a greater or lesser time, according to the kind of life we lead, he goes on to show that there is really no direct evidence that living matter—as such, and *per se*—ever loses its power or vitality, but rather that its power of manifesting is interfered with as life progresses, for the reason that it is forced to occupy a relatively smaller proportion of the whole space of the vital economy, by reason of the clogging and congesting that goes on with the advance of years, and with the altered chemical and physical changes which occur in the organism. Each fragment of "biogen" (living matter) is as powerful as ever, in other words; only it is slowly forced out by the earthy components in the body and compelled to occupy less space. He says, in part:—

". . . Life is never *qualitatively,* but only *quantitatively* diminished; or, in other words, vitality as a physical process never slackens from any variability of its originating force—that force being the universal sentience of matter, and as constant as gravitation and the weight of the earth—and hence death comes to a person, not from a decline of this initial vital power itself, but from those intrinsic obstacles which befall from the material environment and from imperfect modes of living. . . . It is not the sentient constant in 'biogen' that grows old in our ageing organisms, but the surcease of the biogen from the tissues on account of mechanical causes connected with the growth and the product of growth. . . . A tissue is 'old' because there is little biogen in it, not so much because the biogen has grown intrinsically weak."

Mr. Stephens then enumerates the various chemical and physical causes which contribute to old age and death, and points out that all these causes, being understood, might be removed; and that there is no valid reason why death should not be postponed almost indefinitely—looked at from the theoretical point of view.

I have elsewhere dealt with this theory, and will not now discuss it further. His theory of old age contains, assuredly, more than a grain of truth—in fact, is largely true. All the newer researches in cell activity and cell life go to show that *proportions* are changed, but that the *innate power of the proportions* remains practically constant. In other words, living matter is living matter everywhere and always, and its differences are in degree and not in kind. If less of it be present (owing to obstruction or other causes), less of it will be manifest; and if more of it be present, more of it will be manifest.

Few indeed are the men and women of full age—say twenty-five—who have not yet contracted the malady that will kill them, according to that distinguished scientist and physician Dr. Felix Regnault. Normally, as contemporary investigators are beginning to find out, it takes twenty years for a fatal malady to kill a patient. It may take thirty years. The popular impression is that a man may die suddenly, or that he may only require a year to die in, or six months. To be sure, a man may be killed or a child may die in a few months at the age of one year. But ordinarily speaking, all deaths are very slow indeed, and about 95 per cent. of civilized adults are now stricken with fatal diseases. They do not know it. They may not suffer from them. In due time they will have their cases diagnosed as cancer, or as tuberculosis or diabetes, or what not. But so inveterate are current misconceptions of the nature of death that the origin of the fatal malady—in time—will be miscalculated by from ten to thirty years.

In the case of human beings, says Dr. Regnault, writing in *The International* (London), death—barring accident—is nearly always caused by some specific malady. This malady is as likely as not to be cured—that is, what is called "cured." The "cure," however, no matter how skilful the treatment or how slight the disease, has left a weakness behind it in some particular organ of the body. One of the organs is, if not prematurely worn out, at least so worn that its resisting powers are greatly diminished. All of us in this way when we have reached a certain age possess an organ that is much older than the rest of the physique. One day we shall die because of this organ. Even if we live to be very old indeed, we shall not die of "old age" but of weakness of the lungs, or of the kidneys, or of the liver, or of the brain. The individual does not die of senile decay, no matter if he live to be ninety or a hundred. He dies of the decay of the lungs, or of the decay of the heart, or of the decay of the kidneys, or of the decay of some other organ. That organ has been dying for years. For if there be one truth more firmly established than others, it is this: no bodily organ can perish from disease in less than ten years. Sometimes it takes thirty years. Usually it requires twenty years.

How is it that one organ thus decays more quickly than the others? Physicians reply—because it has suffered from the attacks of illness. A cure is never absolute. The organ never comes out of an illness in exactly the same condition as when it went in. Scarlet fever, for example, attacks a person. The kidneys have been thereby affected. For ten, twenty, or even thirty years more they may perform their functions excellently, but nevertheless they will have an earlier senility. The kidney cells slowly perish at a time when the other organs are healthy. At the age of fifty or sixty the sick person is carried off. The same holds true of other and very unimportant illnesses. A man dies of heart-weakness. An old rheumatic attack will very easily be detected as the cause. It long seemed as though it had left no traces, but they show themselves only in the fatal illness. Another old man dies owing to the wearing out of the blood-vessels. If the blood-vessels age more rapidly than the rest of the body, it is because they have been weakened by an infectious disease or some form of poisoning.

Take the case of the man who dies of lung trouble. It is traceable to bronchitis or to slight tuberculosis in youth, which did not betray its presence but yet had weakened the organ. In all cases death is to be ascribed to an illness which had attacked the individual in his youth and weakened an organ, or to some infection which had permanently remained in a latent condition. The bacteria which had caused the illness do not quit the organism when the illness is terminated. They await in the interior of the organ the opportunity for a fresh attack.

"Thus many men who are outwardly healthy carry the malicious enemy inside them. A fever, caught in youth, returns after twenty, thirty, or fifty years; the bacillus, for example, of marsh-fever has been dormant the whole time, and yet in old age awakens to fresh and fatal activity.

"To these causes of the decay of single organs may be added those which are due to the folly of the individual himself. Drinkers ruin their livers, immoderate eaters overload their stomachs, smokers weaken their hearts; life ceases on the day when these organs finally refuse further service. We do not die suddenly; our existence perishes gradually with the weakening of the organs. To reach advanced old age, a man must have been healthy his whole life long."

This theory has been criticized on the ground that it fails to take into account the fact that the body is constantly rebuilding its various parts, particularly its diseased or broken parts, and hence, any innate weakness would be eradicated long before it worked the havoc here suggested. Other reasons, too, might be urged against this theory; but on the whole it is doubtless sound in its main contention, and is a valuable suggestion towards a correct understanding of the causes of death in a large number of cases.

Very different, again, are the views advanced by Dr. Arnold Lorand, of Carlsbad, who has issued an English translation of his work, *Old Age Deferred.* According to this theory, old age and premature death depend, not upon the age of the arteries, as has been so often suggested, but upon the condition of the ductless glands. All vital phenomena, he says, are under the control of the action of these glands; everything depends upon their condition. Symptoms of old age appear after changes in these glands. The appearance, the condition of the tissues, all depend upon their condition. Depressing emotions are, perhaps, the most fatal and certain of all means of breaking down these organs, and insuring premature old age and death. To summarize this author's views in his own words:—

"The symptoms of old age are the result of breakdown of the tissues and organs which, owing to shrinking of the blood-vessels, are insufficiently supplied with blood, and, owing to the disappearance of nervous elements, are devoid of proper nervous control.

"Degeneration of the ductless glands and of the organs and tissues can be simultaneous, for the latter are under the control of the former. These glands govern the processes of metabolism and nutrition of the tissues, and by their incessant antitoxic action protect the organism from the numerous poisonous products, be they of exogenous origin, introduced with air or food, or endogenous, formed as waste products during vital processes. After degeneration of these glands the processes of metabolism in the tissues are diminished, and there is an increase of fibrous tissue at the expense of more highly differentiated structures.

"The fact that the changes in the tissues are secondary and take place only after primary changes in the ductless glands, is best proved by the circumstances that they can be produced, either experimentally by the extirpation of certain of the ductless glands, or spontaneously by the degeneration of these glands in disease.[1]

"It is evident from the above considerations that all hygienic errors, be they errors of diet or any kind of excess, will bring about their own punishment; and that premature old age, or a shortened life, will be the result. In fact, it is

[1] Dr. Voronoff has lately defended this view at length, in his book on *Life*.

mainly our own fault if we become senile at sixty or seventy, and die before ninety or a hundred.

"Not only age, but the majority of diseases, are due to our own fault in undermining our natural immunity against infections, and subjecting our various vital organs to unreasonable overwork and exertion. We do not believe that the worst slave-driver of olden days subjected his slaves to such treatment as we do our own organs, and especially our nerves. At last they must rebel, and disease, with early death or premature old age, will be the result.

"It is literally true, as the German proverb says: 'Jeder is seines Glückes Schmied' (every man is the locksmith of his own happiness), and as a variation on this we would say: 'Every man is the guardian of his own health.'"

Of recent years, Professor Metchnikoff has devoted considerable time and energy to this question of "old age," and discusses the subject fairly and fully in his *Old Age*, mentioned above, in his *New Hygiene*, his *Nature of Man*, and his *Prolongation of Life*. His position throughout all his writings remains the same, and can best be summed up in his own words as follows:—

" . . . I think I am justified in asserting that senile decay is mainly due to the destruction of the higher elements of the organism by macrophags. . . . Since the mechanism of senile atrophy is entirely similar to that of atrophies of microbic or toxic origin, it may be asked whether in old age there may not be some intervention of microbes or their poisons. . . . The principal phenomena of old age depend upon the indirect action of microbes that become collected in our digestive tube. . . . It is really intestinal microbes that are the cause of our senile atrophy. . . . Old age is an infectious chronic disease which is manifested by a degeneration, or an enfeebling of the nobler elements, and by the excessive activity of the macrophags. These modifications cause a disturbance of the equilibrium of the cells composing our body, and set up a struggle within our organism which ends in a precocious ageing and in premature death, contrary to nature."

Accordingly, M. Metchnikoff seeks means to destroy these invading microbes. He thinks that he has found the remedy, in part at least, in the free use of lactic acid, which kills the organisms and renders their growth and presence impossible. Doubtless this method would dispose of the micro-organisms then in the intestinal tube; but what if more are introduced? We must drink more of his soured milk, containing lactic acid! But is it not obvious that this is merely tinkering at effects, instead of going direct to the root and cause of the evil?[1] Why not render the soil such that no microbes can live in it, in the first place, and then no lactic acid treatment or other measures of a similar nature would be necessary? M. Metchnikoff is forced to admit that, if the bowels were perfectly healthy, there would and could be no auto-intoxication, and hence no degeneration of the nature indicated. Why not, then, aim at preserving the bowel in such a state of cleanliness that no micro-organisms could possibly dwell therein? Should not *that* be our ideal?

M. Metchnikoff practically admits this in several passages in his works; but his method of preserving such a state is very different from one that would be recom-

1 Says Professor Charles Minot on this point:—"It is unquestionable that phagocytes do eat up fragments of cells and of tissues, and may even attack whole cells. But to me it seems probable that their *rôle* is entirely secondary. They do not cause the death of cells, but they feed presumably upon cells which are already dead or at least dying. Their activity is to be regarded, so far as the problem of the death of cells is concerned, not as indicating the cause of death, but as a phenomenon for the display of which the death of the cell offers an opportunity."—*Age, Growth and Death,* p. 74. (See Appendix B.)

mended by any hygienic physician. He contends that we should never eat raw food, or food that has not been thoroughly cooked, as we are liable thereby to introduce germs into the intestinal canal! All water should be boiled; everything sterilized—every precaution taken to prevent the introduction into the body of micro-organisms, which he so greatly fears. M. Metchnikoff believes that cancer is produced by micro-organisms, and asserts that he has eaten only cooked foods for many years, in an attempt to escape that terrible malady.

In opposition to this view, I may state that there are many persons—whole colonies of them in California—who eat *nothing but* raw fruits and nuts, and who never boil their water, or cook their food at all—and *they* never suffer from any of these dread complaints, but are, on the contrary, exceptionally healthy and robust and long-lived. Professor Jaffa, who made a special study of these "fruitarians," found them to be especially healthy and possessed of an abundance of energy.[1] And all of these men and women live far longer than the average, and are almost entirely free from the numerous diseases and complaints from which humanity suffers. How is this?

The answer is simple enough. As I have already pointed out in another place, it is not the germ that is to be dreaded, but that condition of the body which renders possible the presence and growth of that germ! If the body were healthy, no germs could live in such an organism, no matter how many were introduced—they would be instantly killed, and they could not exist therein for an instant. We need not bother about the

1 See his *Investigations among Fruitarians*, U. S. Dept. of Agr. Report and my *Natural Food of Man*.

germs; keep the body sound, well, strong, and full of energy, and nature will take care of the rest—including the germs! They are quite incapable of doing any harm in a healthy body. The sounder the body the less danger of infection, and the longer and healthier the life. Now, as fruitarianism, or the practice of living upon fruits, is one of the best possible means of keeping the body in this desirable condition, it will readily be seen that, if we live on raw fruit, and those simple foods that tend to keep the body in the best possible health; and if we are careful, at the same time, not to eat *too much,* we shall keep the intestinal canal free from all obnoxious microbes—for the simple reason that their growth and presence there would be an utter impossibility. No matter if we do introduce into it such micro-organisms with the food, the body would speedily dispose of them. The state of the body is everything; the number of microbes introduced of very small moment. Eat those foods, therefore, that keep the body in the best possible health, and do not worry in the least about the micro-organisms, that may or may not exist in the intestines. They will soon be disposed of. The food is the all-important factor; and fruit—man's natural food—should be eaten almost exclusively if we wish to avoid old age, premature death, and all the ills that exist before it. It will thus be seen that I have been forced to agree with Drs. Bostwick and Evans, previously mentioned, as this was their contention precisely. M. Metchnikoff failed to make sufficient allowance for the germicidal and antiseptic properties of the body, *when maintained in the best of health by means of natural, uncooked foods.*[1] He studied the effects of these

[1] See my *Natural Food of Man* for a defence of this system of diet.

micro-organisms upon bodies badly nourished with cooked food, and food more or less in excess. Had he studied bodies nourished and maintained by their natural food—fruits and nuts, in their uncooked, primitive form—there can be no doubt that M. Metchnikoff would have been forced to the conclusion that, after all, these states are caused by the running down of the vital forces in consequence of the altered chemical condition of the body, and of its blockage by mal-assimilated food-material!

CHAPTER IX

THEORIES AS TO THE NATURE OF DEATH

It has often been said that we cannot know what death is until we know something of the nature of life, and that is very probably true. Death has also been defined, many times, as the "cessation of life." Dr. Minot, e. g., in his *Age, Growth and Death,* (pp. 214–15) says:—

"Death is not a universal accompaniment of life. In many of the lower organisms death does not occur, so far as we at present know, as a natural and necessary result of life. Death with them is purely the result of an accident, some external cause. Our existing science leads us, therefore, to the conception that natural death has been acquired during the process of evolution of living organisms. Why should it have been acquired? It is due to differentiation; when the cells acquire the additional faculty of passing beyond the simple stage to the more complicated organization, they lose some of their vitality, some of their power of growth, some of their possibilities of perpetuation; and as the organization in the process of evolution becomes higher and higher, the necessity for change becomes more and more imperative. But it involves the end. Differentiation leads, as its inevitable conclusion, to death. Death is the price we are obliged to pay for our organization, for the differentiation which exists in us. Is it too high a price? To that organization we are indebted for the great array of faculties with which we are endowed. To it we are indebted for the means of appreciating the sort of world, the kind of universe, in which we are placed. . . . It does not seem to me too much for us to pay. We accept the price. . . . Death of the whole comes, as we now know, whenever some essential part of the body gives way—sometimes one, sometimes another; perhaps the brain, perhaps the heart, perhaps one of the other internal organs may be the first in which the change of cytomorphosis goes so far that it can no longer perform its share of work, and, failing, brings about the failure of the whole. This is the scientific view of death. It leaves death with all its mystery, with all its sacredness; we are not in the least able at the present time to say what life is—still less, perhaps, what death is. We say of certain things—they are alive; of certain others—they are dead; but what the difference may be, what is essential to those two states, science is utterly unable to tell us at the present time. It is a phenomenon with which we are so familiar that perhaps we do not think enough about it."

Haeckel, again, in his book, *The Wonders of Life,* says:—

"The inquiry into the nature of organic life has shown us that it is, in the ultimate analysis, a chemical process. The 'miracle of life' is in essence nothing but the metabolism of the living matter, or of the plasm. . . . If death is the cessation of life, we must mean by that the cessation of the alternation between the upbuild and the dissolution of the molecules of protoplasm; and as each of the molecules of protoplasm must break up again shortly after its formation, we have, in death, to deal only with the definite cessation of reconstruction in the destroyed plasma-molecules. Hence, a living thing is not finally dead—that is to say, absolutely incompetent to discharge any further vital function—until the whole of its plasma molecules are destroyed. . . . Normal death takes place in all organisms when the limit of the hereditary term of life is reached. . . . As Kassowitz has pointed out, the senility of individuals consists in the inevitable increase in the decay of protoplasm, and the metaplastic parts of the body which this produces. Each metaplasm in the body favours the inactive break-up of protoplasm, and so also the formation of new metaplasms. The death of the cell follows, because the chemical energy of the plasm gradually falls off from a certain height—the acme of life. The plasm loses more and more the power to replace, by regeneration, the losses it sustains by the vital functions."

Dr. Brouardel, in his *Death and Sudden Death,* (p. 297), defines death as follows:—

"Death supervenes when poisons manufactured in the system, or unwholesome food that has been ingested, can no longer be adequately removed by the kidneys. . . . The individual is, therefore, poisoned, either by his food, or by poisons which are generated within his own body, i. e. auto-intoxication."

Dr. J. H. Kellogg defines death thus:—

"The cause of old age and natural death is the accumulation of waste matters in the body."

Dr. R. T. Trall, in his *Physiology,* p. 203, favoured the idea that death ensues when—

"The solids are so disproportioned to the fluids that the nutritive process can no longer be carried on."

Dr. Rosenbach contends that—

"Death . . . is that condition of organized matter in which all processes of causation have come to such a state of rest that they can no longer be put in motion, since the grouping of the atoms in the molecule has become so firm that the liberation of living force would be associated with a destruction of the molecule" (*Physician versus Bacteriologist,* pp. 82–3).

Sir Benjamin Ward Richardson, in his *Diseases of Modern Life,* pp. 103–4, sums up the causes of death as follows:—

"I have learned that the gradual transformation of the vital organs of the body from advanced age is due to a change in the colloidal matter which forms the organic basis of all living tissues. In its active state this substance is combined with water, by which its activity and flexibility is maintained in whatever organ it is present—brain, nerve, muscle, eye-ball, cartilage, membrane. In course of time, this combination with water is lessened, whereupon the vital tissues become thickened, or, to use the technical term, 'pectous,' by attraction or cohesion, the organic particles are welded more closely together, until, at length, the nervous matter loses its mobility, and the physical inertia is complete."

All these proposed causes of death may be summed up in the two words *poisoning* and *blockage.*

It will be seen from the above that there exists a great divergence of opinion as to the "true nature" of death —i. e. natural death, and not death due to disease, accident or other causes of a like nature. It was in an attempt to answer this question that I sent out, a number of years ago, a "Questionnaire," in the form of a letter, to a number of eminent *savants,* asking them the question: "What do you consider to be the Real Nature of Death?" A number of letters were received, in answer to this question, but none of them could be considered in any way a definite answer to this question, while many of them displayed a frank indifference as to the whole problem! This curious indifference was pointed out, and in fact proved actually to exist by Professor F. C. S. Schiller, who published a statistical inquiry into the subject in the *Proceedings* of the S. P. R., vol. xviii, pp. 416–53. (See also Mr. Joseph Jacobs' little book, *The Dying of Death.*) This indifference is all the more curious when we take into consideration the fact that every one must one day face it; while the relative lack of interest on the part of physicians and scientists generally is puzzling, from another point-of-view. At all events, it is generally conceded that, in order to define "death," we must first of all define "life," and this I shall now endeavour to do, at least tentatively,—and to define death (natural death) as a consequence of this primary definition.

I need not, indeed, give a definition of the essence of life—its very innermost nature—so long as I can give a satisfactory definition of its *phenomena*, its connection with the organism, and the character of its manifestations through it. If "life" were once understood, we might be enabled to see in what its negation consisted. We can conceive the opposite of a thing we know, but not the opposite of a thing we do not. Roughly, then, let me attempt a definition of the phenomenon of life.

As a matter of fact, I have already offered a tentative definition of life in my *Vitality, Fasting and Nutrition*, p. 334. I then said: "And what is life? That, of course, is unknown. But I venture to think that we shall not go far wrong should we conceive it—on its physical side—for of its essence we are quite ignorant—as a species of *vibration*."[1] I then attempted to show

[1] It may be objected that, in thus defining life as a species of vibration, I have not explained it in full. As I have repeatedly said, I have not attempted to do so. All I intend accomplishing here is a definition of the manifestation of life, and a statement of the possible conditions under which it might or might not manifest. That is all that is attempted, in the case of any other energy or quality. For instance, we do not define the essence of light (so to speak) when we say that it is vibration at a certain rate; nor heat, *ditto*. But we know what we mean by the terms very well, and there is never any demand made to define "light" or "heat" more accurately. I feel that it is the same with life. We are dealing with *phenomena* still, and not with *noumena*. My present idea is merely advanced, to be tested, like any other hypothesis.

how this conception might account for the bodily heat noted, and how it was in accordance with the theory of vitality advanced. Here I shall endeavour to extend the idea in another direction. By it we shall, I think, be enabled to account for death.

If the manifestation of life be actually a species of vibration, and life manifests at a certain rate, and at that rate only (or within certain narrow limits), it will be seen that, in order to render impossible the manifestation of life, it would only be necessary to raise or lower the rate of vibration above or below the limits designed by nature as possible for the manifestation of life, in order to render this manifestation impossible. If the rate of vibration be above a certain speed, life (or its physical base or body) would be shattered, and its manifestation become impossbile. On the other hand, if the rate of vibration were to fall below the minimum limit set by nature, then life would lose its hold of the organism, and drift away, no longer able to manifest through that body. Thus, the power of life *may* be supposed to exercise a variable and fluctuating influence over the body; at times manifesting *fully,* as in intense conscious effort; at times unable to influence it as it should, owing to the poisoned and obstructed condition of the nervous system and the tissues: at times, perhaps, enabled to exercise praeternormal powers and functions; at other times, maintaining only a loose and formal connection, as in sleep, or becoming severed altogether, as in death. This vital energy, this power of life, probably acts more or less indirectly upon the nervous system, through some semi-etheric intermediary. It is the power or energy which sets the molecules of the protoplasm into vital activity. Such, in rough outline, is the theory. Now let us apply it in detail to the facts.

First, let us consider what "natural death" would mean on this theory. Assuming that a certain rate of vibration (of the nervous tissue, or of some ethereal medium acting upon nervous tissue) represents the ideal of health, we might suppose that all rates of vibration above or below this would represent, more or less, manifestations of abnormality or disease—mental or physical. A *slight* lessening of the rate of vibration would indicate a lessened amount of vitality—sluggishness, enervation, depletion, and all that goes with these states. On the other hand, an elevation of the rate of vibration would induce undue excitement, excessive stimulation, abnormal passions and emotions, feverish conditions, *et hoc genus omne.* I need not here go into the medical details of this theory, and of its applicability to disease: possibly I shall do so on another occasion. At present I only wish to indicate its applicability and explanatory power, so far as the phenomena of life and of death are concerned. But any medical man will see its applicability and potentialities, if true.

Now, we may suppose that this rate of vibration would be influenced in two ways—by the condition of the body and by the state of the mind. If the body be choked up with débris, and clogged so that life cannot manifest through it, then the rate of vibration will be so lowered that only a very little life, or life of a low order, can become manifest.[1] If, on the other hand,

[1] This might also account for the phenomena of evolution. Professor F. C. S. Schiller wrote some years ago, in fact (*Riddles of the Sphinx*, p. 294): "If the material encasement be coarse and simple, as in the lower organisms, it permits only a little intelligence to permeate through it; if it is delicate and complex, it leaves more pores and exits, as it were, for the manifestation of consciousness." I venture to think this is quite in line with my theory.

the mind be unduly excited; if it be stimulated and raised to a pitch of great emotional or intellectual activity, this would doubtless correspond to an increased rate of vibratory action; and, thus increasing the activity of the organism through which it manifests, by reason of its raising the rate of vibration of its nerve centres, it would produce disastrous consequences in that organism. A *slight* raising or lowering of this rate of vibration would thus produce disease of one character or of another; and when the rate of vibration exceeds a certain limit, then life would be no longer possible at all—at least in that body.

Now applying this theory to the problem before us, I think we have a satisfactory explanation of natural death, as well as of all sudden deaths—deaths due to accident, disease, etc. Let us see if this is not the case.

Take, first, a supposedly typical case of "natural" death. As the result of years of living contrary to the laws of nature (more or less),[1] the body has become enfeebled, the vitality low, the powers sluggish, the chemical composition of the body altered, and the tissues more or less clogged with débris and mal-assimilated food material. Once this process of blockage and decay has begun, it proceeds more or less rapidly, according to the condition of the organism. Thus, I conceive that it would be impossible for any person in perfect health to die; but no one is in perfect health! This process of blockage, then, goes on from day to day,

[1] No one can live *absolutely* according to the laws of nature. That would be an ideal condition—which does not exist as an actuality. However closely we may obey the laws of nature, therefore, so far as they are known to us, we are perverting them every day of our lives to a great or an infinitesimal degree. For this reason, death ultimately comes to us.

until there comes a time when life cannot set the vital machinery in motion. Thus we have the state which I defined as death in my *Vitality, Fasting and Nutrition,* pp. 330–1, viz.: "That condition of the organism which renders no longer possible the transmission or manifestation of vital force through it—which condition is probably a poisoned state of the nervous system—due, in turn, to the whole system becoming poisoned by toxic material absorbed from the blood."

I now think I see a step further than I did when I wrote that passage. I now think I see *how* it is that life is prevented from manifesting through the body. *It is because the rate of vibration at which life can manifest cannot be reached.* In such a body, the minimum rate of vibration at which life can become manifest to us cannot be attained. Its nervous mechanism cannot be set in motion. I should thus, therefore, define natural death, or death from old age:

> IT IS THE INABILITY OF THE LIFE FORCE TO RAISE TO THE REQUISITE RATE OF VIBRATION THE NERVOUS TISSUE UPON WHICH IT ACTS—ITS MANIFESTATION THUS BEING RENDERED IMPOSSIBLE.

We have here, I think, a fairly complete and satisfactory definition of natural death.

I am well aware of the fact that present-day science does not recognize any such separate life as I have postulated; but, on the contrary, asserts that life is the very *product* of the functioning of the body. Some years ago this was held (in the crudest form) to be true of mind also; but this is now all but universally given up, and I feel assured that the present position with regard to life and vitality will shortly have to be given up also. There is no proof whatever that the present conception or interpretation of the facts is correct; all that science has shown in this particular field is that a certain amount of organic tissue change, and a certain amount of life (so to speak) are present at the same time; and it has been by no means proved that one *creates* the other. All that is proved is *coincidence;* not *causation.* Orthodox science claims that the destruction of a certain amount of *matter* (organic) brings into being a certain amount of *life:* on the contrary, I should hold that the manifestation or expenditure of a certain amount of life wastes or displaces a certain amount of organic matter (which is made good by a proportionately small or large amount of food-material). We can take all the facts of physiological science and interpret them in a different manner. Just as one school of psychologists asserts that the waste of the substance of the brain does not actually produce the thought, but is only coincidental with it (or is even caused by it); so we contend that *all* the vital wastes of the body may be looked at from the same standpoint; and that, instead of the food causing the bodily energy, vital energy wastes the bodily tissues by acting upon them; and this loss is invariably made good by a proportionate amount of food. I argued this position at great length in my *Vitality, Fasting and Nutrition,* pp. 225–303, and Dr. Rabagliati also strongly insisted upon this possible alternate interpretation of the observed facts, in his excellent Introduction to my book; and I shall only refer the reader to the text for an elaboration of that idea. Here I need only say (and on this I insist), that this idea of a separate life force is quite *possible,* and is a tenable position and theory, is in accord with all the known facts of experimental physiology, and also enables us to explain many facts which on the ordinary theory we cannot explain. For these reasons, I accept this theory as substantially correct, and shall proceed with the argument as if it were true.

Dr. Charles S. Minot, indeed, when discussing this difficult question, expressed himself as follows:—

"No mechanical explanation or theory of conscious automatism suffices, but a vital force is the only reasonable hypothesis; the nature of that force is, for the present, an entire mystery, and before we can expect to discover it, we must settle what are the phenomena to be explained by it."

And thirty years later Dr. Minot was still able to say:—

"So little have we gained since 1879 in our comprehension of the basic phenomena of living things, that were I to rewrite the abstract in accordance with present knowledge, I should not change it essentially. The vitalistic hypothesis still seems to me scientifically the best" (*Age, Growth and Death,* p. 267).

It may be added that there is a slow but distinct tendency among biologists to revert to some vitalistic hypothesis for an explanation of living matter and its phenomena. (See, e. g., Wilson, *The Cell,* pp. 394, 417, 434, etc.) And I may point out that if this alternative explanation of the facts is a possible one, it throws an entirely new light on many ill-understood historical phenomena. Thus, Sir Benjamin Ward Richardson, in his *Ministry of Health,* writes as follows:—

"What, then, we should abstractly call 'vitality' is universal, and in persistent operation in inanimate matter constituted to be animated. What we call life is the manifestation of this persistent and all-pervading principle of nature in properly organized substance. What we call death, or devitalization, is the reduction of matter to the sway of other forces, which do not destroy it, but which change its mode of motion from the concrete to the diffuse, and, after a time, render it altogether incapable of manifesting vital action until it be recast in vital mould. We are at this moment ignorant of the time when vitality ceases to act on matter that has been vitalized. Presuming that an organism can be arrested in its living in such a manner that its parts shall not be injured to the extent of actual destruction of tissue or to change of organic form, the vital wave seems ever ready to pour into the body again so soon as the conditions for its action are re-established. Thus, in some of my experiments for suspending the condition essential for the visible manifestation of life in cold-blooded animals, I have succeeded in re-establishing the condition under which the vital vibrations will influence after a lapse, not of hours, but even of days; and, for my part, I know no limitation to such re-manifestation."

The extreme suggestiveness of these remarks need hardly be pointed out, bearing as they do upon the possible interpretation of historic cases of "raising the dead." So long as no part of the organic structure be impaired, or rendered useless, there seems to be no valid reason why the bodily vitality and life should not be forced back into the body by some one, who perhaps understands the law of such vital re-adjustment. On the theory that vitality is a separate force which can exist apart from the bodily structure, this could easily be conceived; and it could even be conceived on the materialistic hypothesis—that vital energy is a mere resultant of the total bodily functions. On that theory life is a product, as it were, of such functionings, and is merely arrested or ceases to be, because the organs which brought life into being cease their activity. Once these organs could be re-stimulated into action, life should re-manifest, by all the laws known to us, or, at least, there is no valid objection, to my thinking, to its doing so. *Why* life does not thus re-manifest in bodies whose structure is unimpaired is really a mystery, but it is probably due to the fact that the *adjustment* between the life force and the body is impaired, and in some way interfered with beyond recall. There are many cases on record in which a man has, e. g., read a telegram and dropped dead instantaneously. Surely the bodily conditions before and after such a catastrophe. must differ almost infinitesimally; and yet there is all the difference in the world between that man's condition before and after such a stupendous event! What has taken place? What physiological reason is there for thinking that life cannot be made to re-manifest in such a body? It has always appeared to me that, were the laws of life—its manifestation and vital connection with the body—more thoroughly understood, cases of "raising the dead" might be far more plentiful than they are at present, and they would no longer be considered "miraculous" by the public at large. Surely this must be because the nervous system is involved, and it is because of the shock that death takes place? The insufficiency of the current theories of life and of death is never more plainly illustrated than in cases of this character. On the materialistic theory, why should stoppage of the heart, or its emptying of blood, cause sudden death? And how comes it about that cardiac massage can restore the heart-beat and life, several minutes after a heart has stopped beating, when the man would normally be pronounced quite dead? On the theory outlined above, it seems to me all such facts are readily explained. Cardiac massage, e. g., would restore a certain vibratory action to the system, which would render possible the re-manifestation of life through it. In the case of the heart-failure, the rate of the life-vibration would be either raised or lowered so suddenly and so tremendously, that its manifestation would no longer be possible. Just as light would suddenly jump into invisibility were we suddenly to increase its rate of vibration, and remain invisible indefinitely so long as we retained that rate, just so would life instantly become invisible and intangible, and would cease to function on this plane, where it is visible or sensible to us, were its rate of vibration suddenly raised in the manner suggested.

Having defined natural death, let us see if this theory applies to all the other known facts, and explains them in a satisfactory manner. I think we shall find that it does. I should begin with death from mental causes.

To take a typical case, a man reads a telegram, and drops dead. In such a case, I have only to suppose that the rate of vibration was raised to such a pitch, in consequence of the mental emotion and excitement, that life shattered itself, as it were, and destroyed its physical basis for life-manifestation. Death from excessive heat or excessive cold would be caused by the gradual raising or lowering respectively of the vibratory action of life. Death from sudden physical shock, jar, electric current, lightning, etc., would raise the vibration of life to such a pitch that it would become extinct immediately. Just as light would cease to be light (for us) as soon as the rate of its vibrations passes a certain number per second, so would life vanish (for us) as soon as the rate of its vibrations passed its proportionately fixed and set limit. In both cases there would be apparent annihilation, but in both cases the vibrations would continue to persist unseen, unsensed, and unknown to us. Life might then persist, after the physical destruction of the body; and, in fact, *must* so persist, unless we are prepared to defend the doctrine of immediate annihilation of energy and the complete upsetting of the laws of evolution, progression, continuity, and the conservation of energy. The bearing of all this on psychical research need only be pointed out in this place.

Coming now to death from diseases of various kinds, we find the same theory equally applicable here, as before. In all such cases life would be unable to manifest, for the reason that it would be unable to raise the rate of bodily vibration to the requisite pitch, in order to manifest through it. The condition would be very much the same as in all cases of natural death, but death would come more suddenly, more painfully, and would cut off a number of years from that person's life. But, beyond some differences in detail, the same cause would apply equally in both cases, and would explain them both equally well.

Now let us turn to the most difficult of all—cases of *sudden* death. Take the case of rupture of the aorta. Personally, I could never see why (apart from the shock to the nervous mechanism) rupture of the aorta, and even its complete emptying of blood, should induce sudden death. The nerves throughout the body are still nourished with blood, and would be for some minutes after the aorta was ruptured. Why, then, should this cause sudden death?[1]

I think all cases of sudden death might be explained on my theory with ease. The sudden raising or lowering of the vibration of life, because of the suddenly induced mental or physical change, would necessitate the raising or the lowering of the vibration of the nervous mechanism accordingly; and, if this were to pass beyond a certain rate in either direction it would mean annihilation—so far as we are concerned—until the rate of vibration could again be lowered or raised sufficiently to allow the re-manifestation of life. In some cases this might be possible. Massage might effect this result, for one thing; hypnotic suggestion or spontaneous trance might bring this to pass, for another—when the vibration had been too high previously—this enabling us to perceive the *rationale* of Miss Molly Fancher's remark, e. g., that "only the trances and spasms saved my life."[2] Electricity might stimulate into action nerves whose vibration had been a trifle too slow to allow of the manifestation of life; other stimulants might act in a similar manner. Only in those cases in which the shock and the rate of change had been so great, that (1) either

[1] In looking through works on sudden death, one cannot help but be struck by the total lack of inquiry into the real cause of such deaths. The author is invariably content to state that death has resulted from a rupture of the cardiac muscles, e. g., or from *angina pectoris*. But when we stop to ask *how* can either of these conditions actually cause death, we find no answer whatever—not even an attempt at an answer! We are told that death does take place suddenly because of these accidents, and that is all. Certainly one cannot be blamed for attempting to formulate a hypothesis which will explain these facts, and look at this matter, not from the mere standpoint of an outsider, helpless; but as one attempting to understand the very innermost essence of the phenomena he sees.

[2] *Molly Fancher*, by Judge A. N. Dailly, p. 22.

the structure of the body had been destroyed; or (2) the medium of re-manifestation had been shattered, would re-manifestation and re-vivification become impossible. In such cases life would be severed from the body for good and all.

This theory of life and its connection with the organism also enables us to explain several puzzling facts which have always been stumbling-blocks in physiology, and still are. I refer to *sleep* and to *insanity.* Let us first take sleep. The innermost nature of this process is a mystery to orthodox science; but it becomes intelligible to us on the theory outlined above. We should suppose, from an analogy, that the vibration of life would get shaken and jarred out of its perfect *rhythm* as the result of the day's excitement and activities; and that it would be necessary to induce some state of the body in which these vibrations of life could be again equalized, ready for the next day's activities; and if this period of rest and re-adjustment were not allowed, then the vibrations would become more and more disturbed, more and more inharmonious and unrhythmic, until the connection of the life force with the organism would be totally mal-adjusted, and insanity and death ensue. *Sleep* would be, therefore, a time of rest necessary for the *re-balancing and re-adjustment of the vibrations of life.*

We have just said that continued loss of sleep will induce insanity more readily and more quickly than almost any other cause. It is asserted that ten days is, as a rule, the greatest length of time that any human organism can exist without sleep. *Why* should so short a period prove so disastrous? I think the answer is found easily enough if the theory just outlined were true. For, in this case, we can see that the longer that sleep is suspended or postponed, the more will the vibrations of life be upset, unequalized, and rendered more inharmonious. We have the analogy of *music* to guide us here. We know that we have harmony so long as the vibrations are *equal;* but so soon as they become *un*-equal in time-interval, then we have discord, or merely *noise.* And it would be the same with the human mind. The continued de-equalization of the vibrations of life would tend to induce more and more mental mal-adjustment to the body; and hence an unbalancing of the mind. *Insanity,* therefore, might be defined as a *condition of the mind resulting from the mal-adjustment and unequalizing of the vibration of life.* It would result from the disharmonious *connection* of the mind and body—a fact upon which Andrew Jackson Davis insisted years ago in his *Mental Disorders.*[1] This would enable us to see why it is that, in the majority of cases, persons insane to us here might be perfectly sane as soon as they died (i. e. so soon as they become "spirits" and this connection consequently no longer existed). It would also enable us to understand the beneficial effects of *music* on the insane—which is now receiving so much attention—and the *rationale* of the various "rest-cures" for mentally sick patients. This unequalizing and unbalancing of the vibration of life would enable us to

[1] He says: "Disturbances, therefore, originate neither in the matter of the body, nor primarily in the principles of the soul, but among the *links*, or rather, in the sensitive connections by which both body and soul are compelled to live together . . . " (pp. 147–8). This theory, of course, is not intended to apply to those cases of insanity in which an actual destruction of brain tissue has taken place, but to those cases where none such is found.

account for all such facts very easily, and would show us why it is that no *physical* disturbance is to be found (*post-mortem*) in a large number of insane patients. We shall have to go beyond materialistic science to explain many cases of this character.

CHAPTER X

GENERAL CONCLUSIONS

THE twentieth century, says Professor Fournier D'Albe,[1] "is too busy to occupy itself much with the problems presented by death and what follows it. The man of the world makes his will, insures his life, and dismisses his own death with the scantiest forms of politeness. The churches, once chiefly interested in the ultimate fate of the soul after death, now devote the bulk of their energies to moral instruction and social amelioration. Death is all but dead as an overshadowing doom and an all-absorbing subject of controversy.

"The spectacle of 2,000,000,000 human beings rushing to their doom, with no definite knowledge of what that doom may be, and yet taking life as it comes, happily and merrily enough as a rule, seems strange and almost unaccountable. The spectacle somewhat resembles that inside a prison during the Reign of Terror, when prisoners passed their time in animated and even gay converse, not knowing who would be called out next to be trundled to the scaffold.

"Every year some 40,000,000 human corpses are consigned to the earth. A million tons of human flesh and blood and bone are discarded as of no further service to humanity, to be gradually transformed into other substances and perhaps other forms of life. Meanwhile the human race, in its myriad forms, lives and thrives. . . . The individual perishes, the species survives. . . . "

As Professor F. C. S. Schiller[2] (of Oxford) also says: "Death is a topic on which philosophers have been astonishingly common-place. . . . Spinoza was right in main-

1 *New Light on Immortality*, pp. 1–3.

2 *Humanism, and Other Essays*, pp. 284–86.

taining that there is no subject concerning which the sage thinks less than about death, which, nevertheless is a great pity, for the sage is surely wrong. There is no subject concerning which he, if he is an idealist and has the courage of his opinion, *ought* to think more, and *ought* to have more interesting things to say.

"In partial proof of which let me attempt to arouse him to reflection by pronouncing some old paradoxes about death which will, I think, be germane to our subject:—

"(1) No man ever yet perished without annihilating also the world in which he lived.

"(2) No man ever yet saw another die; but, if he had, he would have witnessed his own annihilation. . . .

"(3) To die is to cut off our connection with our friends; but do they cut us, or we them, or both, or neither?"

As regards (1), reference is here made to the world of his experience, or, as we might perhaps say with still more accuracy, the objective world, in so far as it was assumed to explain his experience: (2) is true because we can never see another's *self;* what we see is the death of the *body,* which is merely a phenomenon—in *our own* world of experience. Death is not the same thing for him who experiences it and for him who witnesses it.

Says Edward Carpenter (*The Drama of Love and Death,* p. 72):

"One cannot help feeling that—whatever collateral drawbacks there may be in death—the experience itself must be enormously interesting. Talk about starting on a journey; what must the longest sea voyage be compared to this one, with its wonderful vista, and visions, and voices calling? . . . There are lots of books on childbirth and the science of parturition, and the best methods of making the transition easy; but when it comes to the end of life and the event corresponding and complementary to birth, there is little except silence and dismay."

Indeed the subject of death is as little studied as it is fascinating and all but insoluble. For, on its psychological side, it presents the great problems of immortality and the persistence of consciousness beyond the grave. And on its physiological side it presents also (as we have seen) phenomena of the greatest interest. Myers defined death as the "irrevocable self-projection of the spirit,"[1] and attempted to show the link of connection with certain psychic phenomena in this life. Doubtless life presents this psychological side: it also presents a purely physiological side and offers problems for solution which cannot be solved in any such offhand manner as many physiologists would lead us to believe.[2] Indeed, the very moment of death is altogether uncertain—so much so, in fact, that Schultze and Virchow (1870) coined the term "necrobiosis" to designate the transition stage between life and death. Often there is no definite time at which life ceases and death begins; but there is a gradual passage from normal life to complete death, which frequently begins to be noticeable during

1 *Human Personality*, vol. ii., p. 524.

2 Thus: "Death . . . is simply the destruction of protoplasm, which would, of course, destroy its properties" (*The Living World*, by H. W. Conn, p. 32). Apart from accident, however, I see no reason for this "destruction," so calmly supposed. Why should it take place? Again, Loeb (*The Dynamics of Living Matter*, p. 223) says: "In man and the higher mammalians death seems to be caused directly or indirectly through micro-organisms or other injuries to vital organs." Surely this can hardly be considered a definition of natural death, which, according to his own earlier statements, does in fact take place. And, in opposition to this view, Professor Wesley Mills (*Animal Physiology*, p. 669) says: "Few animals perish from simple decay leading to a gradual slowing of the vital machinery down to zero,—so to speak; but when death is not due to violence, as it frequently is, it arises from some essential part getting out of gear, either directly or indirectly."

the course of some disease. Death is developed out of life.

And if this be true, might not the reverse be true also? Might not life be developed out of death? Truly death is a tragedy to those who are left; but is it also a tragedy to the one who has solved the great mystery? If "every cloud has a silver lining" might we not take it for granted that this one has too; and that, beyond "the valley of the shadow," there is surely a hill-top upon which the golden rays of the sun fall with ever-quickening glow? Such would assuredly be the outcome of a cheerful and healthy philosophy! As Stevenson said in his *Aes Triplex*—and I cannot do better than conclude in his stirring words:—

"All literature, from Job and Omar Khayyám to Thomas Carlyle or Walt Whitman, is but an attempt to look upon the human state with such largeness of view as shall enable us to rise from the definition of the living to the definition of Life. And our sages give us about the best satisfaction in their power when they say that it is a vapour, or a shadow, or made out of the same stuff with dreams. Philosophy, in its more rigid sense, has been at the same work for ages; and after a myriad bald heads have wagged over the problem, and after a pile of words have been heaped one upon another into dry and cloudy volumes without end, philosophy has the honour of laying before us, with modest pride, her contribution toward the subject, that life is a Permanent Possibility of Sensation. Truly a fine result! A man may very well love beef, or hunting, or a woman; but surely, surely, not a Permanent Possibility of Sensation! . . .

"Even if death catch people like an open pitfall and in mid-career, laying out vast projects and planning monstrous foundations flushed with hope, and their mouths full of boastful language; should they be at once tripped and silenced, is there not something brave and spirited in such a termination? and does not life go down with a better grace, foaming in full body over a precipice, than miserably struggling to its end in sandy deltas? When the Greeks made their fine saying that those whom the gods love die young, I cannot help believing that they had this sort of death also in their eye. For surely, at whatever age it overtakes a man, this is to die young. Death has not been suffered to take so much as an illusion from his heart. In the hot-fit of life, a tip-toe on the highest point of being, he passes at a bound onto the other side. The noise of the mallet and chisel are scarcely quenched, the trumpets are hardly done blowing, when, trailing with him clouds of glory, this happy-starred, full-blooded spirit shoots into the spiritual world."

PART II

PSYCHOLOGICAL

CHAPTER I

MAN'S THEORIES OF IMMORTALITY

THE problem of the perpetuity of existence is one that has been a strong dynamic force in stimulating and shaping the thought of man from the moment when he arose sufficiently from the plane of barbarism to be able to commence to exercise his mental qualities in this direction. In the earlier days of his existence he may have been satisfied with the life that the objective senses knew; but when the character of the great mystery of the origin and destiny of life began to dawn upon him, the element of dissatisfaction invariably took possession of his thoughts. Instinctively he felt that, as he must have come from something; it was quite reasonable to suppose that he was the object of the solicitude of the Invisible Power that had created him, and that it was the purpose of that Power to convey him through the experiences of this world to some higher plane, where life would flow on more smoothly, or even in an unbroken round of bliss.

It is to such longings for the preservation of the "Ego" that we owe the origin of our religious beliefs and systems, for, as Max Müller says, "without such a belief religion is like an arch resting on one pillar."[1] In the course of his intellectual development there comes a time when man, to some extent, rises above the necessity of a belief in immortality; but there can be no doubt but that such a theory of negation is a mental conception that

[1] *Chips from a German Workshop*, i. 45.

cannot be attained except through the process of reason. That is to say, while it is possible for man to argue himself into a belief in almost anything, he is quite as capable of persuading himself that he believes in nothing—either nothing here or nothing in the hereafter; whereas the question, if it is left to the instinct or to the desires of the human soul, inevitably resolves itself into a cry for protection from the annihilation of the grave. As M. Soldi, the eminent archaeologist, has shown, the rudimentary drawings on the practically shapeless monuments that represent our earliest record of man's physical existence clearly interpret a belief in a survival of conscious existence; and even the savage tribes, far though they may be from the pale of civilization, refuse to admit that the human personality that distinguishes one man from another is destroyed by death. Though the body must perish, that something within the body that stands for individual identity still lives on in the faiths and traditions of almost every land and from the furthest days of antiquity. Naturally, some of these notions are exceedingly crude, and some, perhaps, are extremely materialistic. Behind even the darkest and most obscure ideas, however, the star of hope is shining. Behind the most primitive superstitions there is always the theory that death is not the end of conscious being.

So far as we have been enabled to ascertain, ancestor worship was one of the first systematic religious ideas that the human mind was able to formulate. Prior to the acceptance of this system, religion, where it existed at all, was extremely gross in its sentiments. In the beginning it had undoubtedly consisted in the worship of fetiches, and the semi-superstitious notions that followed these cruder manifestations of man's innate dependence upon a superior power mark the first appearance of the theory of survival among primitive people.

Among savage tribes the idea prevailed that the soul, while in one sense independent of the human body, could not entirely depart from it; and it was due to this notion that both the custom of preserving the body of the dead and the practice of eating it originated. If the corpse was preserved the soul would not be required to abandon it entirely, and could re-enter its envelope on the day of resurrection; while the theory of eating the dead was based upon the belief that this assimilation of the flesh by the relatives of the departed was the best sepulchre that could be provided.

Repellent as these notions may seem, it must be admitted that belief in the survival also led to other criminal customs that are even more horrible to contemplate, for the terrible practice of cannibalism, as well as the slaughter of the aged and infirm, was a ceremonial crime that had its origin in this wrong conception of the future life.

It is a long step from the savage beliefs in cannibalism and the lowest form of metempsychosis to the more civilized worship of ancestors to which the Chinese race still adheres; for, while one represented the grossest sentiment of barbarism, the latter introduced the family institution, a social system that was destined to become one of the leading civilizations of antiquity.

While it is true that Confucius did not explicitly teach the doctrine of the immortality of the soul, his avoidance of this subject does not imply a lack of belief in survival, for there is no record of any time when the Chinese have not believed that, at the moment of death, each person "returned to his family." Even Confucius taught that the spirits of the good were permitted to revisit their former habitations on earth, or such other places as might be prepared by their descendants who desired to pay them homage and receive their benedictions. From this idea came the duty of performing sacred rites in such places, the penalty of any neglect of this service being the loss, to those living, of the supreme felicity flowing from the homage of their own descendants when they, too, had departed. While the survival of the Chinese is in one respect an impersonal immortality, it being a blending of the individual spirit in a kind of collective family-soul, the union of this soul with its descendants is so close that it may almost be said to owe its very existence to the continuance of the homage paid to it.

Among the Egyptians we find the idea of immortality assuming a more definite shape, for they clearly recognized both a dwelling-place of the dead and an actual judgment, with its separation of the just and the unjust. Osiris was to sit as judge, and, all hearts having been weighed in his scales of justice, the wicked were sent to the regions of darkness, while the elect were admitted to a participation in the blissful existence enjoyed by the god of light. Bound up with this very clear idea of immortality were many esoteric doctrines regarding the nature of the soul, as well as beliefs that made the preservation of the body so necessary to the proper continuance of the soul life, that vast tombs were built and the remains of the dead were embalmed, undoubtedly with the intention of making them last for ever.

If it is not easy to find an affirmation of the doctrine of the survival in the books of the Old Testament, it appears in the esoteric books of the Hebrews, and in no uncertain tone. In fact, there can be no question as to the Jewish acceptance of this idea, for while Moses concealed the knowledge that he must have derived from the Egyptians, there is good reason to believe that it was preached to the initiates. In speaking of this reserve upon the part of the illustrious legislator, Bishop Warburton held that this very silence was an indication of his divine mission. "Moses," he said, "being sustained in his legislation and government by immediate divine authority, had not the same necessity that other teachers have for a recourse to threatenings and punishments drawn from the future world, in order to enforce obedience."

Professor Ernst Stahelin, in *The Foundations of Our Faith*, argues along similar lines:—

> "Moses and Confucius did not expressly teach the immortality of the soul; nay, they seemed purposely to avoid entering upon the subject: *they simply took it for granted.* Thus Moses spoke of the tree of life in Paradise, of which if the man took he should live for ever, and called God the God of Abraham, Isaac, and Jacob, thus implying their continued existence, since God could not be a God of the dead, but only of the living; and Confucius, while in some respects avoiding all mention of future things, nevertheless enjoined honours to be paid to departed spirits (thus assuming their life after death) as one of the chief duties of a religious man."

Another evidence that the Jews believed in immortality may be drawn from the laws which Moses promulgated against necromancy, or the invocation of the dead.

This magic art was very generally practised by the Canaanites, and, notwithstanding these laws, prevailed among the Jews at the time of King Saul (1 Sam. xxviii.), or even later (Ps. cvi. 28, etc.).

Job, the Maccabees, and several other biblical books, present a striking exception to this rule of silence regarding the future life which so generally prevails throughout the Old Testament. To illustrate, in Job (xix. 25–27) we may read:—

> "For I know that my Redeemer liveth, and in the last day I shall rise out of the earth, and I shall be clothed again with my skin, and in my flesh shall I see God: whom I myself shall see and my eyes shall behold, and not another; this my hope is laid up in my bosom."

This, as well as many other passages that might be quoted if necessary, show very clearly that the ancient Hebrews not only believed in the survival of the soul, but in a literal bodily resurrection as well, while the *Cabala* and the *Zohar*, the two books that summarize the doctrine taught to the initiates, made it impossible to doubt that the idea of immortality was early adopted by the Jewish people.

All the Hindu sects have a distinct leaning towards the mystic or metaphysical view of life. There are so many different expressions of opinion among Hindu philosophers, however, that it is impossible to select any single theory as one that can be presented as a standard of comparison with other religions. In every instance, however, some sort of survival is recognized, and, in most cases, both the theory of the transmigration of souls and the doctrine of the desirability of a life of purely meditative asceticism are presented.

"To the modern Hindu mind, the soul is a complex creation, made up of a number of fluid-like, invisible elements centred about an immaterial principle. Each of these elements corresponds to a particular faculty of the soul, and may, therefore, be considered as relatively independent of the others. The element is more subtile and attenuated in proportion as the corresponding faculty is higher and more characteristic of man. At death the astral body, accompanied by the superior elements, detaches itself from the physical body, which is now deprived of vitality; it thus preserves a complete individuality, which, according as it is good or bad, determines which place it shall henceforth inhabit as the consequence of its terrestrial existence" (Elbé, vi. 73, 74).

The more abstract conceptions of Nirvana and Moksha have been largely superseded by these and other more modern theories.[1]

Although our knowledge regarding the teachings of the Magi of Chaldea is very incomplete, the portions that have been recovered are quite sufficient to establish the fact that the Chaldeans not only accepted the idea of survival, but that their interpretation of this theory was more rational than the notions displayed by most other ancient nations. The Egyptians, for example, were utterly unable to escape from the idolatrous notions that obtrude themselves into almost every phase of their religious system, whereas the Chaldeans lost all idea of idolatry in their construction of a religion of pure ideals and lofty conceptions. The ancient Chaldeans believed in the survival of the soul, even accepting the idea of a bodily resurrection. As Pausanias says (Book IV. c. xxii.), the Magi always taught that those who had lived pure and just lives would go to the bright realms of

[1] See my *Higher Psychical Development* (Yoga Philosophy) for a detailed account of these beliefs of the Hindus.

Ormuzd, while the wicked would pass into the region of darkness. This doctrine, which was taught by Zoroaster, is still held by the modern Parsees. Their position respecting the soul and its destiny was thus explained by Edward Barucha, a Parsee priest in Bombay, in a communication to the Religious Congress:—

"The one undying spiritual element was created before the body and both were united at birth and are parted at death. The soul, which comes from the spirit world, is possessed of various senses and faculties; it enters the new-born body, out of which it will return at death into the spiritual world. Zoroaster teaches us that God grants to the soul such means and assistance as are requisite for the carrying out of its allotted task: these are knowledge, wisdom, judgment, thought, action, free-will, religious conscience, a guardian angel or beneficent genius, and, above all, revelation. At the resurrection of the dead, when all things shall be renewed and the whole of creation will begin over again, the souls will be provided with new bodies, that they may taste, in the life to come, bliss ineffable."

Among the ancient nations that believed in the soul's survival, none had a firmer or more active faith in immortality than the Gauls; for it was this doctrine of a future life, with its rewards and punishments, that laid the foundation for their institutions as well as their individual life. The acts of heroism with which they accentuated their devotion to the State and their contempt for death, were the direct effect of their belief in a future world. Unfortunately, as in the case of the Druidic and some other doctrines, a complete idea of the teachings of the Gauls' philosophy is unobtainable. Thus we know that they held that man's immaterial part was a divine emanation, and that this was the one vital principle of life. Prior to its appearance as the soul of man, however, it had animated many forms of inferior life—first plants, and afterwards animals. After this experience it was imprisoned in the

"circle of the abyss, *anufu,* but, after long years of struggle and waiting, it escaped thence, and entered the circle of liberty, *abred,* which is also the circle of transmigration. This circle includes all the worlds of trial and atonement peopled by mankind; and of these worlds the earth is one. After many transmigrations the soul will pass on, and will attain the circle of happy worlds and felicity, *gwynfid.* But even this is not all. Far higher and inaccessibly removed is the circle of the infinite, *ceugant,* encompassing the other circles and belonging to God alone" (Elbé, c. viii. 89, 90).

While the Gauls taught the passage of the soul through many forms, their doctrine of transmigration was infinitely nobler than the more or less crude ideas that appear in the early theories of metempsychosis. It was a passing through many bodies, including those of animals, or even plants, but its progress was marked by a steady ascent towards the heights of infinite perfection, and there was no place in this plan for the return of the soul to lower conditions. So thoroughly were the Gauls convinced of the truth of their doctrines, and so firm was their faith in the glory of the future life, that they always gave a condemned prisoner five years in which to prepare for death, not only that he might have time for repentance for his own sake, but for the reason that they did not desire to sully the spirit world by sending a guilty soul into it.

In the Druidical doctrine, the earth was an inferior world devised as a transient abode for the soul during its work of preparation for admission to the world of love. This goal, which is attained only after many transmigrations, is the reward bestowed upon those who have conquered the three great shortcomings of life: (1) neglect of self-instruction, (2) lack of love of good, and (3) attachment to evil.

When the ancient Greeks maintained the idea of survival, as may be seen by an examination of their legendary tales or mythology, it seldom appeared to form a reason for their acts. It was at the foundation of their mysteries, and the arguments that their philosophers advanced for a belief in a future existence are often adapted to modern use. So far as the character of this doctrine of immortality was concerned, however, it was not to be compared to the clearly defined notions maintained by the Gauls. According to the Greek idea, the soul of the deceased person enjoyed at least a semi-conscious existence, in which it retained a sort of half-sensible dependence upon the physical comforts of life. Accordingly the smell of blood of animals, or their cooking flesh, was supposed to be most agreeable to the shades of the dead heroes. It was due to this belief that funeral banquets were held, to which—when the holy fire had been kindled upon the altar of Zeus by the head of the family—the souls of the ancestors were summoned that they might derive their pleasure from the food that was to be sacrificed for the satisfaction of their ghostly appetite.

Crude as these ideas of the people may seem, the poets and philosophers upheld more spiritual theories, by which they not only taught the future existence of the soul, but the two alternatives of good and evil awaiting it after death. Thus Hesiod wrote:—

"Wrapped in the fluid-like envelopes rendering them invisible, the souls of the righteous wander over the earth wielding their regal powers. They mark the good and the evil deeds, and they extend their special protection to such as they have loved in life. As to the souls of the wicked, they are held in Tartarus, where they are punished by the ever-present memory of the crimes which they have committed."

Some six centuries later these views were more definitely and, in many respects, more rationally formulated by Pythagoras, one of the greatest of the world's philosophers. He asserted that, in addition to the natural body, a spiritual element existed—an element possessing unity and surrounded by a semi-material soul. In appearance this soul resembled the body, to which it was so necessary that life would become extinct the moment it was withdrawn. Thus death was the withdrawal of the soul from the body, and in the act of withdrawing it took with it the spirit, or the immaterial element which it enfolded, and proceeded to a region in space corresponding in character to the nature of the deeds that it had performed in the flesh. As Plato described it, the pure soul soared upward with the spirit to the spheres divine, while the impure soul fell back "into the dark regions of matter." To explain the inequality of human conditions and the apparent injustices of life, Pythagoras took refuge in the doctrine of reincarnation.

The Romans were not dissimilar from the Chinese in their early adoption of the system of ancestor worship, and it was upon this religious idea that they constructed the family organization that contributed to the successful upbuilding of their social state. Their ideas of immortality, however, while more impersonal in their tendency than those of the Greeks, were still sufficiently apparent to be recognizable. They did not theorize as to the effect of this future life upon the general harmony of the universe, or apply its rewards to the acts of this existence, for, as Elbé has said, "the thought of immortality appears rather as a pious longing of the imagination devoid of sufficient support in the reality of fact."

Despite this, however, the idea of immortality appears quite conspicuously in the works of many Roman writers. Thus Ovid not only explicitly announces his belief in a future existence, but even adopts the theory of transmigration as a logical explanation of the phenomena of natural life. "Nothing perishes," he says, "but everything changes here on earth. Souls come and go unceasingly in visible form; the animals that succeed in acquiring goodness take upon them human form." Cicero, too, expresses his belief in immortality, and adds that it has been the universal theory from the day of man's first appearance upon earth. To quote the passage from *Scipio's Dream:*—

> "Know that it is not thou, but thy body alone which is mortal. The individual in his entirety resides in the soul, and not in the outward form. Learn, then, that thou art a god; thou, the immortal intelligence which gives movement to a perishable body, just as the eternal God animates an incorruptible body."

As the speculations of Christian theology are described in the chapter on "The Theological Aspect of Death and Immortality," it is only necessary to refer at this time to two ideas that are in direct opposition to these modern religious opinions. These are the ideas of Spiritualism and Theosophy, which embody many of the more or less esoteric doctrines of antiquity, but reproduced in modern form.

Of course, the use of the word "Spiritualism" in this connection is due entirely to the fact that the term has derived the authority of popular approval. Literally, the title "Spiritualism" should be used solely to describe theories that are contrary to those of "Materialism," and in this respect every professing Christian is spiritualistic in his beliefs. Ordinarily, therefore, the term "Spiritism" is much to be preferred; but in this instance we will follow the line of least resistance, and speak of the doctrine of "Spiritualism," as it is generally understood.

It is the teaching of this theory that the discarnate soul, on entering the future world, carries with it the *perisprit,* or astral body, which it had possessed during the period of earthly existence. So far as rewards and punishments are concerned, the soul finds its future already written into the record of its earthly acts. If its mind has been centred upon elevating thoughts, if it has not been too deeply absorbed in material things, and if it has lived in accordance with the purest law of love, it finds it possible to go far from the earth plane, into the condition in which good and righteous souls abide. If, on the other hand, it passes into the next life under evil conditions, it is practically chained to earth. Its perisprit is far more material, and its ability to retain the memory of the pleasures and needs of the physical life inspires so strong a craving for these material things, that it remains close to earth, where it may seize upon every opportunity to appear to the living. When spirits appear under noxious conditions they become what may be termed "evil spirits," or what are popularly known as "demons." It is upon this theory that the idea of demoniac possession is based.

Under more favourable conditions, however, the spirit succeeds in animating the partially free perisprit of a living person, after which it is able to produce the phenomena that have played so important a part in the development of modern spiritualism, sometimes giving communications that are intended to establish the fact of the existence of the personality after death. Dr. Alfred Russel Wallace, in his *Miracles and Modern Spiritualism,* pp. 115–16, thus sums up the belief of the average spiritualist on this question:—

"After death man's spirit survises in an ethereal body, gifted with new powers, but mentally and morally the same individual as when clothed in flesh. Then he commences from that moment a course of apparently endless progression, which is rapid just in proportion as his mental and moral faculties have been exercised and cultivated while on earth. Thus his comparative happiness or misery depend entirely on himself. Just in proportion as his higher human faculties have taken part in all his pleasures, here he will find himself contented and happy in a state of existence in which they will have the fullest exercise; while he who has depended more on the body than on the mind for his pleasures will, when the body is no more, feel a grievous want, and must slowly and painfully develop his intellectual and moral nature until its exercise shall become easy and pleasurable. Neither punishments nor rewards are meted out by an External power, but each one's condition is the natural and inevitable sequence of his condition here. He starts again from the level of the moral and intellectual development to which he had raised himself while on earth."

Emma Hardinge Brittain, in her address on *Hades,* thus further explains the position of the spiritualist:—

"Of the nature of these spheres and their inhabitants, we have spoken from the knowledge of the spirits,—dwellers still in Hades. Would you receive some immediate definition of your own condition, and learn how you shall dwell, and what your garments shall be, what your mansion, scenery, likeness, occupation? Turn your eyes within, and ask what you have learned, and what you have done in this, the school-house for the spheres of spirit-land. There—there is an aristocracy, and even royal rank in varying degree, but the aristocracy is one of merit, and the royalty of soul. It is only the truly wise who govern, and, as the wiser soul is he that is best, as the truest wisdom is the highest love, so the royalty of soul is truth and love. And within the spirit-world all knowledge of this earth, all forms of science, all revelations of art, all mysteries of space, must be understood. The exalted soul that is then all ready for his departure to a higher state than Hades, must know all that earth can teach, and have practised all that heaven requires. The spirit never quits the spheres of earth until he is fully possessed of all the life and knowledge of this planet and its spheres. And though the progress may be here commenced, and not one jot of what you learn or think or strive for here is lost, yet all achievements must be ultimated there, and no soul can wing its flight to that which you call, in view of its earth perfection, Heaven, till you have passed through Earth and Hades, and stand ready in your fully completed pilgrimage, to enter on the new and unspeakable glories of the celestial realms beyond."

Theosophy, on the other hand, is a more mystical philosophy, as it seeks to solve the problems of life, death, and the future existence by means of a system of higher metaphysics. In many cases it has adopted the ideas of the most ancient religions, especially the esoteric doctrines of the Hindu philosophers.

Theosophy declares that the material world is but an insignificant part of the created universe, and that the human being, so far from being confined to a physical body, possesses a spiritual body, or invisible, fluid-like, intermediary body through which the conscious Ego acts. Moreover, this inner body is extremely complex in its construction, being composed of several distinct and different bodies, one encased within another. As summarized by Elbé, these bodies are distinguished as follows:— First, in order of materiality, there is the *etheric body,* which assumes the form and existence of the physical body, to which it is bound by an indissoluble bond. It is composed of ether-like particles that are so infinitely minute that it is impossible to compare them to any earthly substance. Born at the inception of organic life, and expiring at its death, it governs its manifold operations.

Secondly, is the *kamic* or *astral body*, the organ of man's passions and desires. It is the vehicle of feeling and emotion, and through its operations the human being becomes conscious of pleasure, pain, passion, desire, and regret. Although composed of elements that are more subtile than those of the etheric body, the materiality or this body differs in individuals, just as sensitiveness does.

Third, comes the *mental body,* which is the organ of the intellect, and so, of course, manifests itself variously in different individuals.

Fourth, is the *causal body,* through which man conceives abstract ideas, receives the unconscious residue of past experiences, and from which springs the germ that is to expand into future existences.

Lastly, the *Buddhic body*, which is believed to be in a very embryonic state, even in persons of a high degree of righteousness. It is the organ of unselfish love, charity, and self-sacrifice.

While the etheric body, like the physical body, does not survive death, the soul continues to exist in the astral body for a brief or lengthy period, according to its earthly acts. This astral body is finally destined to die, whereupon the soul joyously departs from the plane of conscious suffering in which till then it has found its existence, to ascend, now clothed in the mental body, to a plane of purer ideas and greater bliss. Still even this is but a temporary heaven, or plane of observation, from which the soul can look back and study the various lives through which it has passed, thus viewing the connection existing between the successive existences, and appreciating the happy and unhappy incidents of life in their proper light as the manifestation of the operations of the law of karma, which leaves neither act nor thought unpunished or unrewarded.

If, during the course of these lives, the soul has succeeded in cancelling its debt to karma, and has developed the qualities that compose the Buddhic body, it ascends into yet another world, much closer to God, in which the process of evolution may be continued upon a plane where subsequent reincarnations are unnecessary. But if, in this ascent, the demands of karma have been left unsatisfied, and the thoughts and deeds of life have not been expiated, the soul turns back from this temporary heaven, to pursue its life on earth once more. It is at this time that the most important purpose of the causal body is developed, for it is through the operations of this organ that the various bodies needed as a covering for the immaterial soul are reconstructed. Thus, step by step, the soul that is condemned to live again, enwraps itself in its different envelopes, and, in this manner, it finally shuts out all recollection of the past lives because of which it has been judged unworthy to ascend to the higher realms of light and joy.

While these are by no means the only opinions that man has held upon these ever vital subjects of contemplation, they are sufficient to indicate that he has so far failed to solve the mystery that envelops the fact of being. In spite of all his speculation, there is but one fact that he has been able to establish to his own satisfaction. He is *here,* but whence he came, and whither he is going, or why, are questions to which faith alone has made answer.

Myths as to the Origin of Death.—Many curious beliefs have been held by savage nations as to the cause of death —how death originally came into the world. Tylor, in his *Primitive Culture,* tells us that natural deaths are by many tribes regarded as supernatural. These tribes have no conception of death as the inevitable; as the eventful obstruction and cessation of the powers of the bodily machine; the stopping of the pulses and processes of life by violence or decay or disease. The savage believes that the only real death is due to accident, or by bewitching the unfortunate patient. He knows nothing of "natural" death. For him, man would never die at all, if he were not bewitched, or if some unfortunate accident did not carry him off. Many races in Australia hold this view. The negroes in Central Africa have very much the same belief. Every man who dies what we call a natural death is really killed by witches! The Esquimaux hold similar views.

Myths as to the origin of death are numerous. Usually, death is supposed to have come into the world, owing to some sin of omission (not commission). It was due to the fact that some message from a deity was not properly delivered, or because of the failure to live up to a compact with the gods. Here are some of the Australian myths. "The first created man and woman were told not to go near a certain tree in which a bat lived. One day, however, the woman was gathering firewood, and she went near the tree. The bat flew away, and after that came death." Here is another: "The child of the first man was wounded. If his parents could have healed him, death would never have entered the world. They failed. Death came." Some of the natives of Bengal believe that Death came into the world owing to one of their number having bathed in a certain pool of water, which was forbidden. The Greek origin of death is too well known to need restatement. Pandora and her box will always live in the memory of lovers of art.

In New Zealand it is believed that death came because of the neglect of a ritual process. The Bushman story of the origin of death is very quaint: "The mother of the little hare was lying dead (but we do not know how she came to die). The moon then struck the little hare on the lip, cutting it open, and saying, 'Cry loudly, for your mother will not return, as *I* do, but is quite dead.'" There are several variations of this myth. Some natives believe that death is caused by a snake stealing away souls while God is asleep; in another version, a woman offered to instruct two men *how* to sleep. "She held the nostrils of one, and he never woke at all." In still other cases, death was due to direct murder, in the first instance. In Banks Island it is believed that death came in order to keep down the population, which had become too numerous, owing to man's inherent immortality!

According to the *Satapatha Brahmana,* death was made, like the gods and other creatures, by a being named Prajapati. "Now, of Prajapati, half was mortal, half was immortal. With this mortal half he feared death, and concealed himself from death in earth and water. Death said to the gods: 'What hath become of him who created us?' They answered: 'Fearing thee, hath he entered the earth.' The gods on Prajapati now freed themselves from the dominion of death by celebrating an enormous number of sacrifices. Death was chagrined at their escape from the 'nets and clubs' which he carried in the Aitareya Brahmana. 'As you have escaped me, so will men also escape,' he grumbled. The gods appeased him by the promise that, *in the body,* no man henceforth should evade death. 'Every one who becomes immortal shall do so by first parting with his body.' " (See also Appendix C.)

CHAPTER II

THE PHILOSOPHICAL ASPECT OF DEATH AND IMMORTALITY

THE late Professor William A. Hammond, of Cornell University, writing upon "Immortality," asserted that "the question of the . . . survival of the soul is not a scientific problem. Positive science is impotent either to prove or disprove the dogma," and it is this theory that has been maintained by most philosophers. As Stahelin [1] says:—

"We might take up a line of argument used by philosophy both in ancient and modern times—from Socrates down to Fichte—to prove the immortality of the inner being, an argument derived from the assertion that the soul, being a unity, is, as such, incapable of decay, it being only in the case of the complex that a falling to pieces, or a dissolution, is conceivable. . . . But the abstruse nature of this method leads us to renounce a line of argument from which, we freely confess, we expect little profitable result. For, after all, what absolute proof have we of this unity of the soul? Can we subject it to the microscope, or the scalpel, as we can the visible and tangible? It must content us for the present simply to indicate that the instinct and consciousness of immortality have nothing to fear from the most searching examination of the reason, but find far more confirmation and additional proof than of contradiction in the profoundest thinking. Further, that this instinct and consciousness do actually exist, and are traceable through all the stages and ramifications of the human race . . . is confirmed to us by our op-

1 *Foundations of our Faith*, p. 232.

ponents themselves . . . that there is in man something which is deeper and stronger than the maxims of a self-invented philosophy, namely, the divinely created nobility of his nature, the inherent breath of life, breathed into him by God, the relation to the Eternal, which secures to him eternity."

Watson goes still further, even to the extent of declaring [1] that no where else but in the Bible is there any

"indubitable declaration of man's immortality . . . any facts or principles so obvious as to enable us confidently to infer it. All observation lies directly against the doctrine of man's immortality. He *dies*, and the probabilities of a future life, which have been established upon the unequal distributions of rewards and punishments in this life, and the capacities of the human soul, are a presumptive evidence that have been adduced . . . only by those to whom the doctrine has been transmitted by tradition, and who were therefore in possession of the *idea;* and even then to have any effectual force of persuasion, they must be built upon antecedent principles furnished only by the revelations contained in holy Scriptures. Hence some of the wisest heathens, who were not wholly unaided in their speculations on these subjects by the reflected light of these revelations, confessed themselves unable to come to any satisfactory conclusion. The doubts of Socrates, who expressed himself the most hopefully of any on the subject of a future life, are well known; and Cicero, who occasionally expàtiated with so much eloquence on this topic, shows, by the sceptical expressions which he throws in, that his belief was by no means confirmed."

The first, and, parenthetically, one of the most logical attempts to formulate a philosophical tenet on the doctrine of immortality, is that which is contained in Plato's *Phaedo.* It was upon this presentation that the Neo-Platonists reared their argumentative structure, and

[1] *Institutes*, vol. ii, p. 2.

nearly all the efforts that have been made to find a logical solution to this problem, since that work was written, have been adapted from it.

The Platonic argument for the immortality of the soul may be summarily stated as follows:—(1) The fact that the mind brings to the study of truth a body of interpretive principles and axioms with it, as part of its native endowment, shows that they can be only reminiscential, and, therefore, derived from a pre-existent state; (2) the soul is an ultimate unity—(i. e. monadic in character), and, therefore, not being composite or divisible, it cannot be disintegrated; (3) the term "soul" means the "principle of life," having the ideal of life essentially immanent in it, and inseparable from it, and therefore it must exclude the opposite idea, death; (4) the soul is self-moving, deriving its activity from within; consequently its motion, and, therefore, its life, must be perpetual; (5) the soul as an immaterial reality is essentially related to the immaterial, invisible, eternal idea; and, as the former is akin to the latter in nature, so is it also akin in duration; (6) the superior dignity and value of the soul argue for its survival of the crass body, and even the crass body persists for a time; (7) the cyclical movement of nature shows everywhere the maintenance of life by opposition, as night, day; sleeping, waking; the dying seed, the germinating flower (this is an argument from analogy; out of the decay and death of one living organism, a new life is generated); (8) the instinctive aspiration of the soul towards a future existence shows that the belief is founded on natural law; (9) things that are destructible are destroyed by their peculiar evil or disease; the peculiar evil of the soul is vice, which corrupts the soul's nature, but does not destroy its existence; (10) the world as a moral and natural world demands a future life of rewards and punishments for the rectification of inequalities in this life, else the wrong would ultimately triumph, as in a bad play. This argument is based on the ethical claim that there must be a final equivalence between inner worth and external condition or reward. The views of the Greeks, and especially the views of Plato, have had a profound, an incalculable influence on Christian thought, on early theological formulae, and on the sum of occidental philosophy.

The discussions of the dogma of immortality, which attracted so much attention during the eighteenth and nineteenth centuries, brought no more satisfactory answer to this riddle of existence. In fact, most philosophical writers, who kept within the bounds of logic, came to Emerson's admitted conclusion that "we cannot prove our faith by syllogisms."[5] The French materialists, for example, denied absolutely the possibility of the presence of a soul and the existence of a future life, psychic life to them being purely an organic function. Only less materialistic is the position of the pantheists, headed by Spinoza, for they held that the World-Soul, which, according to their theories, produces and fills the universe, also fills and rules man; that it is only in him that it reaches its special end—which is self-consciousness—and attains to thought and will, but they hold that, at the death of the individual, this World-Soul retreats from him, just as the setting sun seems to draw back its rays into itself; so self-consciousness sinks once more into the great, unconscious, undistinguished spirit-ocean of the whole.

In its effect Schopenhauer's doctrine is not dissimilar.

To him life was the manifestation of the Will-to-Live, and death, the extinction of that Will. In Fichte's system of Idealism, the creative Ego is not the individual, but the Absolute Ego. "The individual Ego realizes itself only by negating its individuality, by universalizing itself, and the Ego thus exemplifying the conceptual life of truth, continues to all eternity, as an indestructible part of the reality of the absolute Ego." So far as individual existence after death is concerned, this is practically the absorption of the Indian Nirvana. Hegel, personally, paid little attention to the problem of life and death, but his disciples were split into two badly divided factions upon this question of continued existence.

Lotze, in his teleological idealism, bases his theory of the immortality of the soul on the principle of value, taking the ground that a thing will continue for ever which by reason of its excellence should be an abiding constitutive part of the Cosmical Order. In other words, immortality, in his opinion, depended upon individual excellence, and was not the fate of all souls. This idea of what may be termed "conditional immortality," was taught by M'Connell, in *The Evolution of Immortality;* by Dr. Edward White, in *Conditional Immortality, or Life in Christ;* and later, by Professor Henry Drummond, in *Natural Law in the Spiritual World,* a work that excited so much attention that more than one hundred thousand copies were sold.

The conditionalist argues that the soul of man has no inherent right to immortality, but that this privilege has been acquired through the operation of the infinite merits of Jesus Christ, who, by his triumph over death, opened the door to a future existence, that pure spirits might participate with him in the joys of eternal life. The sinner who rejects divine grace, however, is doomed to disappear, like all useless organisms, which, in the struggle for life, fail to adapt themselves to existing conditions.

Kant, Locke, and other metaphysicians, agree with those theologians who exclude all these problems as being beyond the province of actual scientific demonstration. They hold that it is impossible to prove a future existence from a belief in a Creator, regardless of the attributes that we may admit that such a Creator possesses. As Professor Hammond indicates, however, in the article already quoted, they admit that "the work of man as a moral being, with infinite potentialities [i. e. infinite possibilities], necessitates an infinite time for their realization." The laws of the moral life are drawn from a transcendental sphere, free from conditions of time and space, and so the very essence of man's moral being is invested with the eternal. Man is infinitely progressive and perfectible in his moral and intellectual evolution, and this fact points indubitably to a further existence. If death were the end, the moral idea would be illusory, and man would perish a fragment. An infinite moral imperative implies an infinite moral ability. Duty demands moral perfection. Further, the moral ideal is a character-ideal, an ideal of personal aim, which implies a personal destiny, and the non-illusoriness of the moral life implies the possibility of realizing its ideal.

Professor Chase, in his article in the *Bibliotheca Sacra,* February 1849, assumed a somewhat similar position, although he expressed the conclusions in a different way, for he based his argument chiefly upon the gradual and progressive development of life in this planet, and this development, in his opinion, when taken in connection with the capacities and endowments of the soul, indicated, on the part of the Creator, a purpose to continue it in being.

Generally speaking, however, the conclusions of most philosophical speculations regarding man's destiny arrive at the same melancholy *finale,* that, whatever we may accept upon faith, or however strongly we may hope for a continuance of existence in another world, there are no *facts* to demonstrate that the tomb does not write the word "Finis" to the book of conscious life. It was such an idea as this that Lord Bacon had in mind when he wrote:—

> "Our inquiries about the nature of the soul must be bound over at last to religion, for otherwise they still lie open to many errors; for, since the substance of the soul was not deduced from the mass of heaven and earth, but immediately from God, how can the knowledge of the reasonable soul be derived from philosophy?"

And even Alger confesses:—

> "The majestic theme of our immortality allures yet baffles us. No fleshly implement of logic or cunning tact of brain can reach the solution. That secret lies in a tissueless realm, whereof no nerve can report beforehand. We must wait a little. Soon we shall grasp and guess no more, but grasp and know."

Although scarcely intended to do so by the author of *The Critical History of the Doctrine of a Future Life,* in the light of modern investigations the words ring with all the promise of prophecy.

CHAPTER III

THE THEOLOGICAL ASPECT OF DEATH AND IMMORTALITY[1]

CICERO defined death as "the departure of the mind from the body," and, if the term "soul" should be substituted for the word "mind," this definition would give a very accurate impression of the theological view of the phyical aspect of this common phenomenon. Thus, Tertullian describes death as "the disunion of the body and soul," and, unsatisfactory as this definition may be in many respects, it is quite as explanatory as most of the conclusions that have been reached by philosophers and scientists. Of course it is easy to see that, even granting the existence of the soul, death would not consist in this separation between the material and the spiritual parts of our being. Such a separation would occur, but it might be the consequence of the death of the body, and not the cause of it. To say that death is the "termination" of life, therefore, is to parry the question. Spencer's "cessation of life" definition is not more evasive.

So far as theology is concerned, however, it has not better terms in which to describe the termination of life, towards which every human being is tending; so it satisfies itself by accepting this single general conclusion, and presenting several theories in explanation of the

[1] A very good summary of what may be considered the theological aspect of death is to be found in Mr. H. M. Alden's volume, *A Study of Death.* The author makes death and sin synonymous terms, and uses them in that manner throughout his work. His book is, consequently, of no use to a scientific writer, and has only historical and religious interest.

phenomena that it cannot adequately define. Thus, since the days of St. Augustine, accepted orthodox theology has held that as sin and death came into the world through Adam's violation of the commands of God, it was not until the second Adam—Jesus Christ—came that the penalty of the first man's disobedience was provisionally forgiven and the birthright of immortality restored to man. That this is the literal teaching of the Bible there can be no question, nor was it questioned to any considerable degree in its application to either animal or man, until the time that the discoveries of geology demonstrated the prevalence of death in ages long anterior to the creation of man, or countless ages before the appearance of sin, as described in the Book of Genesis. The earth's strata are full of the remains of extinct life—life that existed, died, and was buried by the slow process of nature during periods that greatly antedated the appearance of any civilized race. Even before primitive man had left a mark to indicate his occupancy of the earth there was life, some of which had already become extinct, and it is easy to determine, from an examination of these fossil remains, that in those periods life was inevitably followed by, and in many instances actually sustained by, death.

As the result, most theologians now admit that, long before the period of man's innocence, the phenomena of death had their place in the economy of the world. Even then the revolving years were marked by the opening and closing of the earth's foliage; by the ripening, consummation, and decay of the earth's fruits. When our first parents went to drink of the waters of the streams in Paradise, every draught they took to quench their thirst required the destruction of myriads of animalculae, just as the drinking of water does today. If they walked in the fields, or plucked fruits or vegetables to gratify the demands of hunger, each act brought death to some creature that had hitherto experienced the joy of living. In fact theology, as represented by most theologians, now agrees with the assumption of science, that this state of things has existed since the earliest days of the creation of life, and that, in fact, death is the logical ultimate of the law of life under which God, in His good pleasure, placed all creatures that He made, with the single exception of man.

In the case of man, the last and highest creation of the Divine will, the idea of death was immediately set before him as the consequence, or, in fact, the just desert that must follow his disobedience of the law of his Maker, and there is every reason to believe that his very clear conception of the import of this threatening evil was derived directly from the ever-apparent evidence of death's dominion over the beasts of the field, the fowls of the air, and all things that came up out of the earth.

With regard to the animals, or to creatures possessing what may be termed "mere instinct," theology finds nothing that indicates that there is anything judicial in their ordination to death. It is man that has been punished in this manner, and yet, for some reason that is often unexplained, the curse from which he has suffered has also "been brought upon the ground," with the result that "the whole creation groaneth and travaileth in pain together." Just how this arrangement of affairs is to be reconciled with the idea of a benevolent creator is one of those problems which many theologians have found it difficult to solve, but one explanation is given by M'Clintock and Strong:—

"It may relieve the mystery that, as a general rule, the enjoyments of the inferior creatures greatly exceed their sufferings, and death is but little, if at all, the object of their fear, or even a cause of much pain. That 'the sum of animal enjoyment quenched in death is amply compensated by the law of increase and succession, which both perpetuates life and preserves it with the vigour of its powers and the freshness of its joys is certain'; also (as bearing on the physical and moral condition of man, to whom, as chief in this lower world, all arrangements and disposals affecting the lower forms of life were subordinated) their subjection to death has enlarged immensely the extent of man's physical resources, and multiplied manifold the means of his mental development and discipline."

Theology holds that, as "it is appointed unto all men once to die," death is actually a physical necessity devised by the Creator as a means of carrying out His purposes regarding the welfare of the human race, and, being such, it has become a universal law which now extends to all organizations in the material universe. This is an opinion that has long been held by exponents of nearly all schools of theological thought.

At the same time, as it would be contrary to the doctrine of God's omnipotence to urge that such a law must operate despite His desire to maintain the fulfilment of the law of life, it is held that there are other orders of creation dwelling on an immortal plane who are not subject to this condition, the ultimate fate of every human being. It is also held that these creatures are constituted of some kind of material, or, in other words, that so far from being all spirit, they have their own form of organized existence. And in evidence of this, we are pointed to the fact that the bodies of the risen saints were "clothed with incorruption and immortality."

Theology also contends that even these frail bodies of ours, in the antediluvian period, were able to prolong the objective existence to the verge of a millennium, and it is argued from this that it is quite possible for God to imbue the human organism with the power, or the means, of repairing the waste of the forces of life in such manner as to preserve man in unabated vigour and freshness, even to the end of time. It was undoubtedly this belief in God's power to find a means to suspend all laws of His own creation that gave rise to the legend of the Wandering Jew, which was so commonly accepted throughout the entire civilized world a few centuries ago. According to this story, the Jew was punished for his insult to Jesus by being condemned to travel ceaselessly until the Christ should come again in glory on "the last day."

According to the covenant that was orginally given to man, he was to remain exempt from the operation of this law of death so long as he remained obedient to the divine command: "Thou shalt not eat of the tree of knowledge of good and evil, for in the day thou eatest thereof thou shalt surely die." Again, after this law or command had been violated, the Bible is equally explicit in ascribing the beginning of the reign of death over mankind to the transgression of this law: "By one man sin entered into the world, and death by sin; and so death passed upon all men, for that all have sinned."

Drawing its conclusions from such passages of Scripture, theology assumes that immortality was actually originally ordained for man, but that it was only provisionally ordained. Death was the penalty that would be imposed for the violation of the covenant. As long as man remained steadfast to his agreement with God, the latter would abide by the conditions of His sacramental pledge; and it was due to man's trangression of the law that he was compelled to pay the price of his sin by renouncing the gift of immortality that had been promised to him and to his offspring.

In regard to other forms of creation, however, there is no indication in these passage of Scripture that they had also been included in the provisions of this contract. It was man only to whom the law applied, and it is argued, therefore, that the other orders of creatures that may have lived in that time or in preceding stages of the world's existence, were exempt, both from the necessity of obedience to the law and from the penalty required in case of its violation.

Whether, in any way, they may have been constituted under a law of death by anticipation, and as in keeping with a state of things in which death should reign over man, we do not venture to pronounce. That, indirectly, as a consequence of their relation to man as a sinner against God, their sufferings have been increased and their lives shortened, it is impossible to doubt or deny. But if, in this view, sin be the occasion of their death, it cannot be the cause of it. They are incapable of sin, and cannot die judicially for sin. The contrary opinion which long and generally prevailed, that the creatures were immortal until man sinned, has as little to justify it in Scripture as in Science. Death, it is there said, is the law of their being; and the true doctrine of the Scripture is not that they die because man has sinned, but that man because he has sinned has forfeited his original and high distinction, and has become "like the beasts that perish." It is unnecessary to multiply Scripture proofs of this awful and humbling truth. Every one is familiar with the frequent and equivalent testimonies that death is "the fruit," "the wages," "the end" and consummation of sin, and the circumstances which attend and induce it impressively connect it with sin as its cause.

In order to argue that death, now that it has come, does not necessarily mean the end of all life; theologians have assumed that, in addition to the objective body, each human being possesses a soul or spirit. In this opinion, of course, they differ from the materialist, who holds that man is composed of a physical body alone; that, in fact, he is no more that a superior animal, whose mental and moral strength are merely the effect of the higher development of this physical organism. To the materialist, the theory that man is possessed of a soul distinct from the body, and that it is this soul that is the seat of the nobler intelligence, is the height of absurdity. To the theologian, however, this belief is a necessity, for if it were not for the existence of this spiritual part of man, it would be impossible to show that death, instead of being the end of all things, is really a second birth—a birth into another, a more important, and an eternal state of being.

As evidence of the existence of the soul, the theologian points to numerous passages of Scripture, for the books of the New Testament are filled with testimony that conclusively establishes the truth of this theory, provided we are willing to accept them as authoritative. According to these passages, man's intercourse with the outward existence is through the body, which is entirely objective in its mode of operation, but his communion with God and his ability to attain any degree of spiritual development comes to him through the soul, or spiritual self that, while associated with the body, is a distinct and different organism.

The effect of this complexity of being not only appears in the affairs of life, but also tends to complicate the nature and the result of death. If man had his body alone, it would be easy to dispose of such a problem, for death extends to every part of the body, and includes every portion of his objective nature. In this manner the threat that death should follow the violation of the divine command has been literally enforced. Man does die or perish, so far as his earthly body is concerned. The important question is in regard to the other self—the soul, the spirit, through which he, in accordance with the Biblical promise, may eventually experience far greater joys of living.

In reading the Bible it is easy to discover that reference is made to two kinds of death—the death of the body and the death of the soul—or a spiritual death as well as a material death. In other words, while condemning the outward or objective man to experience "the pangs of death" as a punishment for his sins, God does not permit the inner man—the actual cause of that sin—to escape the penalty. On the contrary, the Scripture assures us that the soul that sinneth shall surely die, and there are many references that might be made to passages that indicate that it is possible to be dead in sin while yet alive in the objective or physical body.

Precisely what effect God's penalty has upon the body and soul, both severally and together, constitutes questions over which there has been considerable dispute.

According to some theologians, it is the actual physical body that is to be raised from the grave on the "last day." In the opinion of others, the physical body will play no part in this resurrection, but, being dead, will perish for all time, while the soul alone will be called upon to account for the sins committed in the flesh.

Naturally, the literal effect of death upon the bodily organism is a matter of common observation. When death comes, the body soon loses its comeliness. Corruption follows, and finally, the structure that was once a human form becomes a shapeless mass of dust. That this dust should be brought together again, to serve once more as the soul's envelope during eternity, is an idea that is branded as absurd by nearly all so-called rational thinkers. In spite of its apparent absurdity, however, this theory has been very generally held by nearly all schools of theology, and it is still accepted by many Christian sects, some of which could scarcely be designated as "primitive."

Whatever disposition may be made of the body, all creeds admit the eventual immortality of at least a certain number of the souls of those who have died. Just when this eternity of bliss is to open its doors to the waiting soul, or to what degree divine mercy will operate in extending the scope of the plan of redemption, are among the many questions that are still in dispute. One school, more liberal than others, grants eventual salvation to all mankind; another school—the Roman Catholic Church—provides an intermediate state, or place of purification, in which those who do not merit eternal damnation may expiate the sins committed in this life; while the several schools of Protestant theology differ in their conceptions of the plans of divine justice—ranging from the ultimate salvation of all, as preached by the most liberal Christians, to the final and absolute extinction of the wicked, a doctrine that seems to have lost many of its adherents during the past few years.

Unpopular as this belief may have become, Hudson, in his *Law of Psychic Phenomena,* insists that it is the only logical view of the situation. He says:—

"The first proposition of my theory is that the death or practical extinction of the soul as a conscious entity is the necessary result of unbelief in immortality. The second proposition is that the soul, having attained immortality through belief, is then subject to the law of rewards and punishments, 'according to the deeds done in the body.' The same propositions are more sententiously expressed in Romans ii. 12: 'For as many as have sinned without law shall also perish without law; and as many as have sinned in the law shall be judged by the law.'

"In other words, the condition precedent to the attainment of immortality, or salvation—that is, the saving of the soul from death—is *belief.* The condition precedent to the attainment of eternal bliss, and the avoidance of the punishments incident to sin, is righteousness.

"Again, we find a spiritual penalty following a violation of spiritual law in what Christ taught regarding the sin against the Holy Ghost. Just what that sin consists in, never has been satisfactorily defined. We are told that it is a sin that cannot be forgiven. It must therefore consist of a violation of some fundamental law of the soul's existence, the penalty for which is inevitable according to the fixed laws of God. It cannot be a moral offence, consisting simply in wrong-doing, for such sins can be atoned for. . . . It must, therefore, be the sin of unbelief, and consist of a blasphemous denial of the soul and its Father, God. This would be in strict accordance with the fundamental law of suggestion."

So far as the popular view of death and life after death is concerned, it now differs widely from the position that theology must take when it decides to stand by the logical aspect of the question. According to the popular impression, the souls of those who die go directly to the seat of judgment, and remain eternally in the realm of bliss if they are able to establish their worthiness to participate in this glorious existence of the blessed. That this idea is suggested by preachers and teachers of religion there can be no doubt, but it is equally certain that the Bible nowhere presents any such theory. On the contrary, it teaches that, as sin and death came into the world through Adam, it was through Christ, the sacrifice, that they have been overcome. It was Christ alone who came sinless into the world, and who lived a sinless life; it was Christ alone who arose triumphant over death, and it is through the acceptance of Christ alone that the soul can be saved. This is the promise of the Bible: that he who believes in Christ, and who lives in accord with that faith, shall have eternal life, but no salvation is promised for those who fail to take advantage of this opportunity. In fact, for the unbeliever, the best that is offered is eternal darkness. To him the eternity that is to be so blissful an experience for the faithful Christian becomes a burden, a period of ceaseless torment—either of mind or body—a time of "weeping, and wailing, and gnashing of teeth."

This, in brief, is the position to which the logical student of theology must turn, contradictory though it may be to the popular view upon these questions, for there can be no doubt that this is the literal gospel that is taught in the Bible. The sin of Adam, which separted him from God, and which sent him flying with fear from the Garden of Eden, began its disrupting work at the very moment of his trangression. The act of treason—the violation of the covenant—intercepted the happy intercourse that had existed between God the Maker, and man His creature. In this manner God's contract was instantly fulfilled. Man had sinned, and God ceased to live with him. The law of God had been broken, the fruit of the forbidden tree had been eaten, and death, through this sin, was brought into the life of the man—the one creature of earth to whom the provisional promise of immortality had been given. "In the day thou eatest thou shalt surely die," God had said, and in the day that he ate the work of death began in the creature's disrupted relations with the Creator.

Although these conditions are sufficient to cause death (and death without delay), through the mercy of God, as displayed in the plan of atonement, man lives on in the body for a little time, that he may have an opportunity to take advantage of the new covenant that God has made with him. Through the expiation of the Cross the doors of eternal life have been opened once more, that he who will may enter. The price that must be paid is *faith* plus *works,* and to every man is given the chance to win this prize which was once lost through Adam's sin: immortality—not immortality in this life indeed, but in a life that lies beyond the grave. Thus, whatever the result of God's forbearance and mercy to each individual soul, the physical man must still die. In theology this mortal crisis which each and all must face is known as the "temporal" death, to distinguish it from the "spiritual" death, which makes it possible for a man to be "dead while he liveth."

When this point has been reached, when this dread day has come, theology recognizes but one more step before the complete and final issue is attained, for when the last plans of the divine administration have been realized, and the God who created all things is ready to take His own unto Himself, the bodies of all who have slept in dust will be reorganized; both the just and the unjust shall rise from their graves to stand with the quick before their judge, that they may give an account of their experience in the flesh, and be judged in accordance with their deserts. It is then that the just shall be raised by faith through grace to the life eternal and incorruptible, while the unjust, the unbeliever, and impenitent sinner shall go away from God's presence into the place of everlasting punishment which is devised for the "resurrection of damnation." It is this that is meant by the "second death." Such is the "orthodox" theological conception of death and "immortality." Naturally the degree to which such teachings influence us will depend upon the extent of our belief in the authenticity of the Scriptures themselves. From the strictly scientific point of view, these teachings will remain theories only, and science will depend upon facts alone for her conclusions.

CHAPTER IV

THE COMMON ARGUMENTS FOR IMMORTALITY

WHILE it is true that positive science has been unable —with microscope or scalpel—to find the smallest trace of an immortal spark in man; while theology has nothing more evidential to offer than an appeal based upon the presumptive accuracy of its revelation; and philosophy stands ready to confess its inability to cope with the problems of death and the continuance of conscious existence, the great majority of human beings are quite as confident of the reality of the next world as it would be possible for them to be if their theories were supported by the most conclusive scientific evidence. Of course, as has been shown, the inability of man to demonstrate the mere existence of the soul to the satisfaction of any law of logic leaves mankind absolutely dependent upon the hope that is in him, that instinctive desire for immortality, the arguments for which are so beautifully summarized by Addison:—

> "Plato, thou reason'st well,
> Else whence this pleasing hope, this fond desire,
> This longing after immortality?
> Or whence this secret dread, and inward horror
> Of falling into naught? Why shrinks the soul
> Back on herself, and startles at destruction?
> 'Tis the divinity that stirs within us;
> 'Tis heaven itself that points out an hereafter,
> And intimates eternity to man."[1]

True as these words may be, if we are to regard them solely as a picture of man's protest against the doctrine

[1] *Cato.*

of extinction, they are not of the faintest evidential value in support of the belief that life continues beyond the grave. In fact, as Hudson said:[1]—

"Natural theology stands precisely where it did when Thales philosophized and Simonides sang; and the arguments are identical with those which Socrates employed in his confutation of the atheism of Aristodemus. Not one of the physical sciences in which we excel the Idumeans has advanced us one step in solution of the great problem propounded by Job, 'If a man die, shall he live again?'"

Professor Hammond mentions five traditional arguments that have commonly been used to establish the fact that death is not the end of conscious being. These are:—

"(1) The ontological argument, which bases immortality on the immateriality, simplicity, and irreducibility of the soul-substance; (2) The teleological argument, which employs the concept of man's destiny and function, his disposition to free himself more and more from the conditions of time and space, and to develop completely his intellectual and moral potentialities, which development is impossible under the conditions of earthly life; (3) The theological argument: the wisdom and justice of God guarantee the self-realization of personal beings whom He has created; (4) The moral argument, i. e. the moral demand for the ultimate equivalence of personal deserts and rewards, which equivalence is not found in life; (5) The historical argument; the fact that the belief is widespread and ancient, showing it to be deep-seated in human nature, and the historical fact of the resurrection of Christ and the statements of the New Testament Scriptures."

As all of these arguments have already been con-

[1] *A Scientific Demonstration of the Future Life*, p. 27.

sidered in previous chapters, it is unnecessary to dwell upon them further, except to the degree in which they apply to the arguments that are in more common use among men; for while many of us may be unable to follow the philosophers and logicians through the intricate mazes of reasoning that lead to their ultimate conclusions, there are certain arguments—more common-place, perhaps—that appeal to ordinary thinkers as extremely convincing. As Hudson says in *A Scientific Demonstration of the Future Life*:—

> "It may sound very unscientific, but I must confess that I attach more of scientific value to Emerson's dogmatic assertion that 'man is to live hereafter' than I do to the aggregate of philosophical speculations known to the literature of the subject. He was one of those pure, lofty, and poetic souls whose intuitive perception and recognition of truth is oftentimes as perfect as a mathematical demonstration."

And there are many individuals who, whether their process of reasoning be scientific or not, will heartily agree with this statement.

Of course, as a matter of fact, there are but two methods of reason that can be applied logically to any question. One is inductive reasoning—the reasoning which begins with accepted facts, or particulars, and from them argues up to the last logical conclusions. The other is deductive reasoning, or the reasoning that begins with conclusions, and from them argues down to facts. Inductive reasoning, therefore, is a logical appeal to fact; whereas deductive reasoning takes the facts that have been obtained more or less inductively, and from them proceeds to calculate its logical particulars. While both methods of reasoning are perfectly legitimate, therefore, for both depend upon the accuracy of the facts, or observations, upon which these conclusions are based, the inductive method is the only scientific one to apply to a problem such as this.

As may easily be imagined, the exponents of the doctrine of life after death have found it extremely difficult to present a very conclusive inductive argument in support of their theories, owing to the absence of facts from which to approach the general conclusion. Accordingly, the tendency shown by modern science—both biology and physiology—has been to dismiss the theory of the soul's existence as undemonstrable.

At the same time it must not be imagined that no attempt has been made to adduce a sound and rational argument based upon the accepted facts of science. Thus, the relations existing between the molecular movements of the brain and their manifestation in human thoughts and feelings have been held to be evidence of the fallacy of the materialistic theory. Professor James, in *Human Immortality,* attempted to "draw the fangs of cerebralistic materialism" by ascribing to the brain a "transmissive" function, but many scientists have not accepted his theory.

It cannot be too strongly insisted upon that, because mental activity and molecular changes always go hand in hand, the one is not therefore *produced* by the other. It is certainly true that for every thought we think, there is a corresponding change in the brain substance; but this merely proves the *coincidence* to us, and does nothing to solve the problem of *causation.* We know that there is a definite activity of the brain during all thinking processes, but that does not tell us what the activity is. It is usually assumed that this is a causative function, but that is merely an assumption, as a matter of fact; and Professor William James and other writers have shown us, and indeed insisted upon the fact, that this function might be other than causal in character—it might be coincidental, or even the result of mental operations! In his *Human Immortality,* Professor James contended that this function of the brain might be a *transmissive* function just as well as a causal one; and, on that theory, consciousness might exist apart from the brain, and merely function *through* it; and such an interpretation of the facts would leave us quite open to believe anything we pleased regarding the possible separate existence of consciousness. At all events, it would appear that there is no valid reason, physiologically considered, for denying immortality; it is merely a question of interpretation of the observed phenomena. Although certain facts would seem to tell in favour of materialism at first glance, it will be seen that this alternate explanation is always open to us; and hence physiology is as helpless as philosophy when it comes to this question of immortality—and the possibility of solving the problem on *a priori* grounds.

In the *Unseen Universe,* by Stewart and Tait, an effort was made to establish the existence of an unseen world from which this world has come, and to which it is connected by bonds of energy. These physicists believe that their theory explains both the origin of molecules and the force which animates them. They claim that the idea that the visible universe has the power to originate life is utterly contradictory to the facts of observation and experiment; and they assert that the hypothesis of an eternal unseen universe is necessary if we are to explain the law of evolution, the conservation of mass and energy, the law of biogenesis (every living being presupposing an antecedent life), the law of continuity (there being no break in reality, the universe being of a piece), and other recognized phenomena of life in the visible world.

Convincing as such arguments may seem to those to whom they appeal, the critical mind is compelled to admit that their validity does not stand the test of the infallible rules to which all such propositions must logically submit. So, too, the analogical argument inevitably falls when exposed to the analysis of the rules of correct reasoning.

In presenting the details of the analogical argument in support of a future life, it is impossible to summarize such speculations more briefly and completely than by quoting from Alger's *Critical History of the Doctrine of a Future Life:—*

"Man, holding his conscious being precious beyond all things, and shrinking with pervasive anxieties from the moment of destined dissolution, looks around through the realms of nature, with thoughtful eye, in search of parallel phenomena further developed, significant sequels in other creatures' fates, whose evolution and fulfilment may haply throw light on his own. With eager vision and heart-prompted imagination, he scrutinizes whatever appears related to his object. Seeing the snake cast its old slough and glide forth renewed, he conceives that in death man but sheds his fleshly exuviae, while the spirit emerges regenerate. He beholds the beetle break from its filthy sepulchre, and commence its summer work; and straightway he hangs a golden scarabaeus in his temples as an emblem of future life. After vegetation's wintry deaths, hailing the returning spring that brings resurrection and life to the graves of the sod, he dreams of some far-off spring of humanity, yet to come, when the frosts of man's untoward doom shall relent, and all the costly seed sown through ages in the great earth-tomb shall shoot up in celestial shapes. On the moaning seashore, weeping some dear friend, he perceives, now ascending in the dawn, the planet which he lately saw declining in the dusk; and he is cheered by the though that—

> " 'So sinks the day-star in the ocean-bed,
> And yet anon repairs his drooping head,
> And tricks his beams, and with new-spangled ore
> Flames in the forehead, of the morning sky:
> So Lycidas sunk low, but mounted high.'

"Some traveller or poet tells him fabulous tales of a bird which, grown aged, fills his nest with spices, and, spontaneously burning, soars from the aromatic fire, rejuvenescent for a thousand years; and he cannot but take the phoenix for a miraculous type of his own soul swinging, free and eternal, from the ashes of his corpse. Having watched the silkworm, as it wove its cocoon and lay down in its oblong grave apparently dead, until at length it struggles forth, glittering with rainbow colours, a winged moth, endowed with new faculties, and living a new life in a new sphere, he conceives that so the human soul may, in the fulness of time, disentangle itself from the imprisoning meshes of this world of larvae, a thing of spirit beauty, to sail through heavenly airs; and henceforth he engraves a butterfly on the tombstone in vivid prophecy of immortality. Thus a moralizing observation of natural similitudes teaches man to hope for an existence beyond death."

Butler, in the *Analogy,*[1] presents a very similar argument, assuming that because the caterpillar is transformed into the butterfly, and "worms into flies," we are to exist hereafter "according to a natural order or appointment of the very same kind with that we have

1 Part I. c. i.

already experienced''; but, like Alger and other exponents of analogical reasoning, he makes the mistake of trying to adapt poetic figures of speech or fanciful comparisons to questions that must be determined upon a purely logical basis. To be legitimate, analogical reasoning must justify itself by its conformity to all the conditions of correct logical induction. Thus the field in which analogical reasoning may be properly employed has very decided limitations. It may be proper to employ it when dealing with matters which are known to be governed by the same or substantially the same laws; but never when instituting comparisons, either between subjects which are known not to be governed by the same laws, or between subjects which are not known to be governed by the same laws. . . . In all inductive reasoning there is one proposition that is, or may be, always assumed, namely, *the constancy of nature.* Thus by the observation of a series of phenomena, say the rising and the setting of the sun, we are enabled to predict with absolute confidence that it will on any given day in the future rise in the east and set in the west. Why? Because we have such confidence in the immutability of the laws of nature that we may assume that the order of the rising and the setting of the sun will never be reversed. It is upon this assumption of the constancy of nature, or rather upon the sublime verity of this assumption, that all advancement in the arts and sciences depends; for if it were not true, we would derive no certain information from our experience, or from our observation of the phenomena of nature. If gravity operated one day and on the next refrained from operating, the whole human race would be instantly put to confusion, and lose faith in the integrity of

Nature. Inductive reasoning, therefore, could have no possible value as a means of interpreting the laws of nature but for the fact that we know that nature is ever constant.

It is interesting to note, however, in this connection, that Professor S. P. Langley did not believe that any "Laws of Nature" exist at all, as a matter of fact, but are merely mental conceptions! Thus he believed that Nature exists, and that her phenomena are uniform, and from this uniformity we have constructed modern science, and formulated what we choose to call "the laws of nature." But these laws do not exist as absolute, fixed realities, as a matter of fact; they are merely mental concepts. At any time new facts may come upon the scene, which will make us alter our conception of the laws of nature, and extend them in a fashion hitherto undreamed of. And yet the laws are not really altered, in the old sense of the term; the fact was that no such laws existed as we had postulated, and constant readjustment must be made to fit new facts. Professor Langley insisted upon this over and over again, and wrote in this connection:—

"The immensely greater number of things we know in almost every department of science beyond those which were known one hundred and fifty years ago has had an effect which doubtless could have been anticipated, but which we may not have wholly expected. It is, that the more we know the more we recognize our ignorance, and the more we have a sense of the mystery of the universe and the limitations of our knowledge. . . . Innumerable are the illusions of custom, but of all these perhaps the cleverest is her knack of persuading us that the miraculous, by simple repetition, ceases to be miraculous. . . . Suppose that a century ago, in the year 1802, certain French academicians, believing like every one else then in the 'laws of nature,' were invited, in the light of the best scientific knowledge of the day, to name the most grotesque and outrageous violation of them which the human mind could conceive. I may suppose them to reply, 'If a cartload of black stones were to tumble out of the blue sky above us before our eyes in this very France, we should call that a violation of the laws of nature, indeed.' Yet the next year, not one, but many cartloads of black stones did tumble out of the blue sky, not in some far-off land, but in France itself.

"It is of interest to ask, what became of the 'laws of nature' after such a terrible blow? The 'laws of nature' were adjusted, and after being enlarged by a little patching, so as to take in the new fact, were found to be just as good as ever. So it is always; when the miracle has happened, then and only then it becomes most clear that it was no miracle at all, and that no 'law of nature' had been broken.

"Applying the parable to ourselves, then, how shall we deal with new facts which are on trial, things perhaps not wholly demonstrated, yet partly plausible? During the very last generation hypnotism was such a violation of natural law. Now it is part of it. What shall we say, again, about telepathy, which seemed so absurd to most of us a dozen years ago? I do not say there is such a thing now, but I would like to take the occasion to express my feeling that Sir William Crookes, as president of the British Association, took the right, as he took the courageous course, in speaking of it in the terms he did. . . . Though nature be external to ourselves, the so-called 'laws of nature' are from within—laws of our own minds—and a simple product of our human nature." [1]

To return, however, we see that analogical reasoning is a form of deductive reasoning, and it depends for its

[1] *Smithsonian Report*, pp. 545–52.

validity upon the accuracy of the facts which it assumes. Thus, when it begins to argue from one subject up to an entirely different subject, or attempts to make conditions existing in one class of phenomena apply to phenomena that are entirely dissimilar in character, it is treading upon dangerous ground. If the laws governing the subject-matter observed are not identical with those of the subject-matter investigated, the analogical argument must fall to the ground from sheer lack of supporting facts.

Professor Chase, in his *Bibliotheca Sacra* article already mentioned, objects to Bishop Butler's argument as being "less fortunate than any other part of that great work:" In particularizing, he said: "Both of the main arguments employed by him are no less applicable to the lower animals than to man, and just as much prove the immortality of the living principle connected with the minutest insect or humblest infusoria as of the human soul. It is not a little remarkable that this fact, which in reality converts the attempted proof in a *reductio ad absurdum* of the principles from which it is drawn, should not have awakened in the cautious mind of Butler a suspicion of their soundness, and led him to seek other means of establishing the truth in question. These he would have found, and, as we think, far better suited to his purpose, in the facts and principles so ably and so fully set forth in his chapters on the moral government of God, and on probation considered as a means of discipline and improvement."

In addition to the particulars to which Professor Chase objects, there are other directions in which these analogical arguments fail to meet the test of criticism. For example, if it had merely been inferred that because a silkworm is metamorphosed into a butterfly, other larvae were destined to be transformed into winged insects, there might be a reasonably logical basis for such an assumption, because the laws governing the one case might reasonably be assumed to be applicable to all other like cases. The laws governing man, and those that apply to the life of insects, are, however, of an entirely different class, and to attempt to make the mortal life of one prove the immortal existence of the other is certainly an illegitimate use of the principles of analogy, especially in view of the fact that the changed conditions of the life of this insect do not in any sense present the elements of immortality, as the insect dies after its transformation is concluded. Equally fallacious are Butler's arguments based upon the hatching of the bird from the egg, or the birth of man from the womb. In no case do the same physical laws act as the governing force. On the contrary, as several writers have said, the presumptions from analogy, when they are legitimate, are against rather than in favour of the continued existence of man after death. If we take Nature as an illustration, her phenomena would lead the logical mind to assume that death is actually the end of the process of life. Even the analogical argument drawn from the germination of the seed fails as ignobly to apply in the case of continued existence, for the vegetable life that is derived from the seed that has fallen to the ground and disintegrated is in no respect the same life as that which existed in the plant from which the seed was produced, and, if the analogy applies to man at all, it simply bears out the theory of the materialistic scientists who hold that man's immortality is in his posterity. If we believed, like the Saracens, that the individual soul is instantaneously transferred to the universal soul at death, there might be some logical justification for the assumption that an analogy exists "between the gathering of the material of which the body of man consists from the vast store of matter in nature and its final restoration to that store, and the emanation of the spirit of man from the universal intellect, the Divinity, and its final reabsorption."[1] As here, also, however, the validity of this analogy depends upon the assumption of the existence of a soul—and this is the very fact that must be proved, not assumed—it would fall far short of meeting all the logical requirements of science. In fact, no analogy can be instituted—

> "Between the operations of physical nature and those of the spiritual realm . . . unless it is first clearly shown that the laws of the two worlds are identical. And as it is manifestly impossible to know the laws which prevail in the unseen universe, it follows that reasoning from such analogies is not only unsatisfactory to the last degree, but, measured by logical and scientific standards, it is, to employ no harsher expression, positively nugatory. It is like trying to demonstrate a proposition in mathematics by citing a rule in grammar. Nor does it avoid the objection to express the analogy in the negative form, which was such a favourite of the late Bishop Butler; for it is the logical equivalent of saying, 'There is no presumption from analogy to be found in the rules of grammar against the possibility of squaring the circle. Therefore the circle can be squared.'"

There are many Christians who feel that it is little better than a waste of time and brain-matter to endeavour to establish the fact of immortality when this

1 Draper's *Conflict between Religion and Science.*

doctrine has been so explicitly taught in the New Testament. It was this analogical argument that Paul used when (1 Cor. xv. 14) he wrote:—

> "Now, if Christ be preached that He rose from the dead, how say some among you that there is no resurrection of the dead?
>
> "But if there be no resurrection of the dead, then is Christ not risen, and if Christ be not risen, then is our preaching vain, and your faith is also vain."

Even should we go to the extent of assuming—as science will not—that the New Testament narrative of the resurrection of Jesus is literally true, and that every doctrine of the Christian faith can be substantiated beyond the shadow of doubt, we are still confronted by the same objections, that exist in every instance in which the argument from analogy is made to apply to the question of immortality. Thus, if the dogmas of orthodox Christianity be true, and Christ the God who was raised from the dead, the very fact of His divinity subjects Him to the operation of a different kind of law from that which governs mankind. Moreover, if we are to assume —with some other sects—that, while Christ was mere man, the Father performed a miracle in restoring Him from death, this assumption leaves us in the same position as before, for what right have we to imagine that because a miracle was performed in this case the law of nature is to be violated for every man who dies?

It will be remembered, of course, that these objections to the New Testament approval of the doctrine of immortality are simply raised to show that we must resort to something more evidential than mere prescriptive authority if we are to prove the continuance of conscious life to the satisfaction of science. As an intuitive argument, the teachings of the various scriptures are of far more importance as showing the persistence of man's belief in a future life.

As has already been shown in previous chapters, the antiquity of this belief in eternal life is beyond question. All races have held it, and in all ages it has been the star of hope to which all men have instinctively turned. Thus, Alger says:—

> "It is obvious that man is endowed at once with foreknowledge of death and with a powerful love of life. It is not a love of being here, for he often loathes the scenes around him. It is a love of self-possessed existence, a love of his own soul in its central consciousness and bounded reality. This is the inseparable element of his very entity. Crowned with free-will, walking on the crest of the world, enfeoffed with individual faculties, served by vassal nature with tributes of various joy, he cannot bear the thought of losing himself or of sliding into the general abyss of matter. His interior consciousness is permeated with a self-preserving instinct, and shudders at every glimpse of danger or hint of death. The soul, pervaded with a guardian instinct of life and seeing death's steady approach to destroy the body, necessitates the conception of an escape into another state of existence. Fancy and reason, thus set at work, speedily construct a thousand theories filled with details. Desire first fathers the thought, and then thought woos belief."

This restless yearning for another world, a realm in which the disembodied spirit may continue the conscious existence that the physical senses now know as life, has been at the bottom of all religious faiths; but while it is possible that this belief in immortality may be the logical expression of the immortal spark within us, this argument, conclusive though it may be to many persons, does not, and will not, satisfy science until we can demonstrate that it is something more than the material instinct of self-preservation which is common to all physical organisms. Like the analogical argument presented by Alger and Butler, the argument of intuition applies to the lower animals quite as logically as it does to man. Though we may look upon ourselves as of more account in the eye of the Creator, the final product of an evolutionary process to which we are no longer subject, such presumptions do not constitute a particularly valid argument. As Schopenhauer says:—

> "Every one feels that he is something different from a being who has once been created from nothing by another being. In this way the assurance rises within him that although death can make an end of his life, it cannot make an end of his existence." [1]

Clearly as Schopenhauer states this argument with which mankind has sought to establish a basis for its belief in immortality, man is too logical a reasoner not to recognize the fact that such a theory cannot be adapted exclusively to the members of the *genus homo.* "Man is something else than an animate nothing," he asserts, and with this all who believe in immortality will agree; but to this he adds, "and the animal also."

In conclusion, I may say that, while I am ready to admit that the presence of the belief in eternal life in almost every human heart may be taken as a *presumption* that such a desire may yet be realized, I still deny that such theories as those described can logically be accepted, as a conclusive argument. Before the doctrine

1. *Indestructibility of our Nature by Death.*

of the continuance of conscious existence after death can be accepted, as *proved,* we must demonstrate that another world actually exists, and that in this unseen realm the disembodied spirit, by whatever name we may designate it, continues to maintain the individuality that it possessed on earth. When this result has been attained, and not until then, will man be justified in regarding his hope for immortality as anything more than the manifestation of that instinct of self-preservation that has ever been the "first law of nature."

CHAPTER V

THE TESTIMONY OF SCIENCE—PSYCHICAL RESEARCH

ALL things perish! So far as we can see, there is not one thing in the universe which escapes that fate, unless it be energy. Science has always contended that every individual organism must die; that, no matter how long death may be postponed, it must come sooner or later. Everything in the universe perishes, it was said—all but two things, matter and energy. But now the newer school of physicists contends that matter, too, perishes, and that the old dogma of the indestructibility of matter is erroneous, and not in accord with the latest discoveries of modern science. Yet, oddly enough, *life,* the most precious of all the energies, is supposed to become extinct at death! The energy we call life is supposed, it is true, to pass into other modes of energy; but it does not persist *as such.* The only trouble experienced by those who condemned this view and contended that the mental life did persist after bodily dissolution was that there was no evidence that it did! In the absence of this proof the doctrine had, naturally, to be given up.

When consciousness came to be treated as a function of the brain, still more doubt was thrown upon the belief, which now seemed to have no solid ground for its rational support. On the materialistic theory, consciousness was considered a mere product of the brain's functioning—a position, however, open to many objections, as

I have elsewhere shown.[1] But in the absence of positive proof to the contrary, there was always a justification for the scepticism that prevailed, for the most part, during the closing years of the last century.

Count Solovovo, of Petrograd, has, indeed, well expressed this view in a letter to the author. He says:—

> "I believe it most probable that *Death* is the end of everything throughout the whole realm of Nature. I believe that everything tends to support this conclusion; everyday experience, scientific experiment, and observation, and last—not least—plain common sense. And, before all, I am convinced of the utter inability of religion to grapple satisfactorily with the problem. And if, in spite of all that, there is still a lingering doubt in my mind that this negative conclusion, though overwhelmingly probable, may yet be not absolutely certain, I owe this shadow of a doubt to certain alleged facts of psychical research, so-called, only and exclusively.
>
> "PEROVSKY-PETROVO-SOLOVOVO."

So we come to psychical phenomena. *Here* are facts which (apparently at least) prove that consciousness *does* persist apart from the body, and evidence is produced in support of that belief. Whether the evidence be sufficiently conclusive or not is, of course, another matter; but no one can dispute the fact that this is the rational *method*—the right way of solving the problem, and the only way in which it can ever be solved. Arguments as to world-theories and metaphysics might go on for ever; but if definite facts can be produced, indicating that some consciousness is active (that consciousness having been severed from its body previously), then rival theories will have to be adjusted to the facts, and only

[1] *The Coming Science*, pp. 114–179; *The Physical Phenomena of Spiritualism*, pp. 413, 414.

by such facts can the question ever be satisfactorily settled.

We have seen, during the course of our discussion of this problem, that there is no *a priori* "impossibility" as to the persistence of life after the destruction of the body; from the standpoint of pure physiology, all the known phenomena can be interpreted just as readily in one way as in another. We have seen, further, that there are certain indications that consciousness can persist unclouded up to the very moment of physical dissolution, when a sudden and wholly incomprehensible change takes place, transforming the living being into a mere "corpse." We say, in popular parlance, that the "life has left the body." Has it in truth done so—or has it become extinct? That is the all-important question to be settled! Let us see if there are not some indications that "something" persists after the destruction of the physical frame to which we have grown only too accustomed to think of as the real "Self."

In the first place, then, certain cases of so-called "Visions of the Dying" seem to indicate some heightened, supernormal power of vision at this time—as though the subject were enabled to peer behind the Veil, if only for a few moments, at this transitional period. Dr. Hyslop has published a number of interesting cases of this character in his *Psychical Research and the Resurrection,* pp. 81–108. Mere hallucinations are common enough, and are, perhaps, only to be expected; but in certain cases, the dying person has stated that he had seen and conversed with "dead" relatives or friends, when the fact of their death had been carefully kept from the patient, on account of the possible "shock" the news might occasion. Yet the subject stated emphatically that these individuals were "spirits"—i. e. were dead and not living—but that they were waiting to see and greet him on the "other side"—as soon as he "passed over." These cases are certainly suggestive, so far as they go.

Again, the process of dying has been observed and minutely described by certain "clairvoyants," and they tell in great detail what happens! They state that the soul is *withdrawn* from the body (taking some time in the operation), and they have narrated precisely what occurs,—as seen by their psychical vision. I have quoted a typical account of this nature in my *Psychical Phenomena and the War*, pp. 229–41, to which the interested reader is referred,—as well as for accounts of the process of dying, as described by *soi-disant* "spirits." Whatever objective value one may attach to such narratives, they certainly agree remarkably, not only with one another, but also with phenomena which may occasionally be observed in our daily life.

These phenomena are the cases of self-projection, "astral projection," in which the soul and body seem to be separated from one another,—for a time at least,—and, while the physical body remains entranced or asleep, the "astral" or "etheric" Double journeys to some distant scene, and is seen and recognized in that place by independent observers. Many such cases are on record, and may be found in *Phantasms of the Living*, the *Proceedings* of the S. P. R., Dr. I. K. Funk's book, *The Psychic Riddle*, Myers' *Human Personality*, and other works of a like nature. Such instances certainly seem to indicate that the *psyche* of man may be detached from his physical body, and may exist independently, in a life of its own, even in this world—and the actual details of such self-projection can—it is said—be acquired.[1]

Again, experiments have been undertaken in an attempt to weigh and photograph the soul at the moment of death. Dr. Baraduc, of Paris, obtained a remarkable series of photographs of this character, which have never been explained, and which were certainly not due to imperfections upon the plates. These photographs—taken every fifteen minutes for three hours after death—show a remarkable and progressive series of luminous clouds, over the body laid-out on the bed. I have reproduced some of these, together with details of the experiments undertaken—in my *Problems of Psychical Research,* to which the reader is referred for additional details.

Experiments in "weighing the soul" were made, some years ago, by Dr. Duncan MacDougal, of Haverill, Mass. The patient was placed—bed and all—upon a delicate balance, and, at the exact moment of death—a sudden loss of from half to three-quarters of an ounce was noticed, which caused the beam to fly upward and strike the upper steel arm with a sharp "click." Four out of six experiments yielded positive results. (I have described these experiments in detail in my book *The Coming Science,* and to this the reader is referred for further particulars.)

These experiments, so far as they go, certainly seem to indicate that *something* leaves the body at the moment of death. What is that something? Has it intelligence and memory, or is it merely a form of matter or energy? The phenomena of "Apparitions of the Dying" seem

1 (I have summarized these teachings and exercises in my *Higher Psychical Development:* chapter, "The Projection of the Astral Body.")

to show that some form of consciousness is certainly connected with this "something," which leaves the body at such times, and that this Being can manifest at great distances from the body,—and perhaps even convey information to the "percipient" which the latter did not know. The phantom seems, then, to carry a certain consciousness with it. But this would be quite impossible upon the materialistic theory, which contends that life and consciousness are inseparably bound-up with the body, and incapable of existing apart from it. If the objective character of these phantoms be proved, then we have evidence that such separate existence is possible; while if the phantom be seen some hours, days or weeks after death, it would certainly indicate that this conscious entity has continued to exist after the death of the physical body—that is, as a "spirit" or "soul," living in some non-material world.[1] For the evidence as to cases of this character, the reader is referred to the *Proceedings* of the S. P. R., to Myers' *Human Personality* and to other authoritative sources of a like nature.

But I shall not press the evidence afforded by these phenomena unduly. Many of the instances do, indeed, seem to indicate in the strongest possible manner that some active entity is present, and that ordinary, normal explanations are totally insufficient to account for the facts. Still, there is the objection that unknown biological and psychological forces within ourselves may account for such phenomena—aided perhaps by telepathy, clairvoyance, etc. For this reason, I shall not place undue emphasis upon such occurrences—suggestive as they are—nor shall I press the evidence afforded by the "physical phenomena" of spiritualism, nor of the su-

1 Phantasms of the Dead, Haunted Houses, etc.

pernormal powers of the subliminal consciousness, so skilfully presented by Myers. All such evidence is, indeed, *confirmatory;* none of it in itself can be said to me *conclusive,* or to offer "proof" of the sort desired by materialistic science.

How then is such "proof" to be obtained? In one way and in one way only. *By proof of personal identity.* Only in that manner can the persistence of consciousness—or what is commonly known as "the immortality of the soul"—be proved. The great objection of materialistic science to such phenomena must always be borne in mind, viz., that consciousness is a function of the brain, that life is bound-up with the material body, and that neither can exist apart from the body and its functionings. The only way in which one can meet this argument is to adduce evidence which tends to show that life and consciousness *do continue* to exist apart from the physical body; and the only way in which this can be done is to adduce evidence of the persistence of memory and personal identity.

The evidence desired in order to prove this, and the *only* evidence that ever will prove it, is the establishment of the identity of the deceased person; and it will be seen that this can only be done by obtaining specific facts and details from that consciousness, which were known to it when alive, but which were presumably in the possession of no *other* consciousness. If we could get in touch, directly or indirectly, with what claimed to be such a consciousness, therefore, and it could produce for us certain facts known only to it when alive (which facts we were enabled afterwards to verify), then we should have fairly good evidence for the belief that such an individual intelligence was operative in the case before us. When once the facts pass beyond the limits of chance, guessing, inference, telepathy, and clairvoyance, and when the honesty of the medium has been proved, there would seem to be no other alternative than to accept the doctrine of "spiritualism," as at least a thinkable and working hypothesis.

It is, of course, quite impossible for me even to summarize the evidence which has been published in the past, in support of this contention. This evidence is often of an extremely complicated character, requiring great care and patience to follow, and covers many thousands of closely-printed pages, scattered through nearly a hundred volumes of the publications of the English and American Societies for Psychical Research, and in hundreds of books now published upon this subject. This evidence has been obtained by means of automatic speech and writing, through numbers of professional and non-professional mediums, and through individuals who do not consider themselves "mediums" at all, in the strict sense of the word. It is doubtless true, however, that some of the very best evidence of personal identity ever obtained has come through Mrs. Piper; and this has been supplemented, of late years, by "messages" coming through Mrs. Leonard, Mrs. Holland, Mrs. Verrall and other mediums. This evidence may be said to fall into four main classifications: (1) Personal Evidence; (2) Classical Knowledge; (3) *Post-Mortem* Letters; and (4) Cross Correspondences or "Concordant Automatisms."

(1) *Personal Evidence.* This material consists in attempts to prove personal identity pure and simple. A "spirit" purports to communicate through an entranced medium, and narrates facts about his (or her) past life,

—which facts may or may not have been known to the sitter, but which, upon investigation, turn out to be true. Very striking evidence of this sort has been obtained—evidence which it has perhaps taken months of correspondence to verify, and which was entirely unknown to the sitter at the time of the sitting. Detailed, personal, trivial incidents constitute, in such cases, the *best* evidence obtainable—since they would be the least likely to have been known to the sitter, and the least likely to have been known by the medium. The following incident is, perhaps, typical of this. It is taken from Dr. Hodgson's Second Report on Mrs. Piper, S. P. R. *Proceedings,* vol. xiii., pp. 321–22:—[1]

Mr. Howard: Tell me something now that you remember that had happened before.

G. P.: Well, I will. About Arthur ought to be a test. How absurd. . . . What does Jim mean? Do you mean our conversations on different things? or do you mean something else?

Mr. Howard: I mean anything. Now, George, listen for a moment; listen, listen.

G. P.: I know.

Mr. Howard: I mean that we spent a great many summers and winters together, and talked on a great many things, and had a great many views in common, went through a great many experiences together. Now (G. P. commencing to write), hold on a minute.

G. P.: You used to talk to me about . . .

The transcription here of the words written by G. P. conveys, of course, no proper impression of the actual circumstances; the inert mass of the upper part of Mrs. Piper's body turned away from the right arm, and sagging down, as it

[1] G. P.—The initials of the supposed "communicator" or spirit, while Mr. Howard is the "sitter." S. P. R.—Society for Psychical Research.

were, limp and lifeless over Mr. Howard's shoulder; but the right arm, and especially hand, mobile, intelligent, deprecatory, then impatient and fierce in the persistence of the writing which followed, contains too much of the personal element in G. P.'s life to be reproduced here. Several statements were read by me and assented to by Mr. Howard, and then was written "private," and the hand gently pushed me away. I retired to the other side of the room, and Mr. Howard took my place close to the hand, where he could read the writing. He did not, of course, read it aloud, and it was too private for my perusal. The hand, as it reached the end of each sheet, tore it off from the block book, and thrust it wildly at Mr. Howard, and then continued writing. The circumstances narrated, Mr. Howard informed me, contained precisely the kind of a test for which he had asked; and he said that he was "perfectly satisfied, perfectly." After this incident there was some further conversation with reference to the past that seemed especially natural, coming from G. P.

In the following remarkable instance, again, the sitter (Dr. Frederick van Eeden) carried on a long conversation with the "communicator" *in Dutch*—a language of which the medium—a middle class English woman—was totally ignorant. In his Report in the *Proceedings* S. P. R., Dr. van Eeden says:—

". . . For instance, the young man who had committed suicide gave as proofs of his identity Dutch names and places which were not at all in my mind at the moment. This might have been unconscious telepathy. At the same time proper names were given which I had never heard myself. Yet later, in Holland, I came across people who bore these very names, though their connection (if any) with the young man I could not find out. . . .

"My personal impression [of the value of the evidence] has varied in the following manner. During the first series of experiments, in November and December 1899, I felt a very strong conviction that the person whose relics I had brought with me, and who had died fifteen years ago, was living as a spirit and was in communication with me through Mrs. Thompson. A number of small particulars, which will be found in the notes, produced on me, when taken *en bloc*, the effect of perfect evidence. To regard these all as guesses made at random seemed absurd: to explain them by telepathy forced and insufficient.

"But when I came home, I found on further inquiry inexplicable faults and failures. If I had really spoken to the dead man, he would never have made these mistakes. And the remarkable feature of it was that all these mistakes were in those very particulars which I had not known myself, and was unable to correct on the spot. . . .

"Consequently, my opinion changed. There were the facts, quite as certain and marvellous as before. I could not ascribe them to fraud or coincidence, but I began to doubt my first impression that I had really dealt with the spirit of a deceased person; and I came to the conclusion that I had dealt only with Mrs. Thompson, who, possessing an unconscious power of information quite beyond our understanding, had acted the ghost, though in perfect good faith. . . .

"But on my second visit, in June 1900, when I took with me the piece of clothing of the young man who had committed suicide, my first impression came back, and with greater force. I was well on my guard, and if I gave hints, it was not unconsciously, but on purpose; and, as will be seen from the notes, the plainest hints were not taken, but the truth came out in the most curious and unexpected ways. . . .

"Up to the sitting of June 7th, all information came through Nellie, Mrs. Thompson's so-called spirit-control. But on that date, the deceased tried, as he had promised, to take the control himself, as the technical term goes. The evidence then became very striking. During a few minutes—though a few minutes only—I felt absolutely as if I were speaking to my friend himself. I spoke Dutch and got immediate and correct answers. The expression of satisfaction and gratification in face and gesture, when we seemed to understand one another, was too vivid to be acted. Quite unexpected Dutch words were pronounced, details were given which were far from my mind, some of which, as that about my father's uncle in a former sitting, I had never known, and found to be true only on inquiry afterwards. . . . "

This is the character of the evidence presented in proof of personal identity. Thousands of pages of similar material have been published, chiefly in the *Proceedings* of the S. P. R. The above quotations are merely intended to illustrate the character of the evidence received, and the methods of attacking this problem adopted by psychical researchers. Let us now turn to the second line of attempted proof.

(2) *Classical Knowledge.* The education of the average individual may usually be gauged with fair accuracy from a few minutes' conversation; certainly any very profound knowledge upon any profound subject would reveal itself should that subject be brought into the conversation. Classical knowledge, now-a-days, is relatively rare among the rank-and-file of humanity—that is to say, profound and exceptional knowledge, which even classical scholars themselves would have to verify. Such knowledge could hardly be possessed by the average medium, or even the exceptional medium. Yet knowledge of just this character *was* obtained through Mrs. Leonard,—purporting to come from F. W. H. Myers and Professor Butcher, of Cambridge,—both of them classical scholars of exceptional erudition. A thorough knowledge of the classics was evidenced by them; frequent references were made to classical authors,

etc.—all of which might be supposed to have been known to the "spirit" authors of these messages, but none of which could conceivably have been known to the medium; and indeed scholars themselves had to ascertain the correctness of many of the references made. A typical instance of this is contained in Prof. G. W. Balfour's paper "The Ear of Dionysius." Evidence of this sort points very strongly, it seems to me, to the genuine character of the messages received, and is indicative of some form of the "spirit hypothesis," rather than to any form of telepathic theory, or to any form of "subconscious activity" on the part of the medium!

(3) *Post-Mortem Letters.* The character of this evidence is as follows: A., when alive, writes a letter; he seals it and sends it to the Society for Psychical Research, where it is deposited in the safe of the Society. A. is now the only living consciousness which knows the contents of this letter. After a time, A. dies. He then purports to communicate through a medium, and says: "I am A. I wrote a certain letter, and in it I said—so-and-so,"—giving the message. The sealed letter is then opened, and the two messages compared. If they are found to correspond, it is pretty good evidence that the *same* consciousness that wrote the one also wrote the other—that is, that the same person is still "there," dictating the contents of the letter. Several such tests have been tried, but it is regrettable that nearly all of them have signally failed. The "thought," in one or two letters, however, has been strikingly similar. Several letters are now on file in the offices of the Society, and while we do not wish their writers any hard luck—still, we are anxious to try the tests! Further evidence in

this direction is greatly to be desired, and will assuredly be most interesting, when obtained.

(4) *Cross Correspondence.* This method was improvised—at the suggestion of the "spirits" themselves—to shut off, so far as possible, the mutual influence of telepathy,—or the influence upon the medium's mind of thoughts contained in the sitter's mind,—telepathically influencing the results. The principle of "cross-correspondence" is as follows: Let us take four mediums, A, B, C and D, all living in different parts of the world, and unknown to one another. Through three of these mediums come parts or fragments of sentences which in themselves are meaningless. Then, through a fourth medium comes a fourth fragment, with the remark: "See the scripts of A, B, and C; these, pieced together with this message, will make a consistent whole." Upon these various scripts being pieced together, a perfectly consistent message is the result. What would that prove? Assuming the facts to be as represented, it would surely tend to prove—*not* that the subconscious minds of these four mediums were telepathically passing-along the information one to another,—since they would not know *what* to pass,—each portion in itself being meaningless. It would tend to prove, rather, that *one* mind was directing and dictating all four messages, through as many mediums—this mind being external to all of them, and itself possessing the key to the puzzle from the beginning. This external mind would be, in short, a "spirit." Such experiments have been conducted on a large scale, and a number of elaborate Reports have been published, describing the experiments in great detail. These "cross correspondences" have, in a

number of instances, been eminently successful, and have gone far towards proving the persistence of consciousness to a number of eminent psychical researchers. The experiments are still continuing.

Such, in brief, is the method which psychical research is following in an endeavour to obtain direct proof of survival. Whatever we may think of the results, we must admit that the *method* is the correct one. In such a work as this, it is of course impossible to do more than refer to the evidence, and to the character of the investigations undertaken; the reader must consult the sources themselves for the actual proofs obtained. While, to many minds, the evidence so far published is insufficient to warrant a belief in the existence of "spirits," it is at least sufficiently strong to show us that here is a case for investigation; one, moreover, which presents great possibilities; and which, if the conclusions were established, would definitely and finally defeat materialism.

CHAPTER VI

CONCLUSION

In the foregoing work, I have endeavoured to present a fairly complete study of Death from the physiological, historic and psychological view-points. To what extent I have done so must be left to the reader—and the future!—to decide. We have studied the causes of old age and the phenomena surrounding Death, and we have endeavoured to discover the innermost nature of the facts thus observed. Further, we have extended our vision, and tried to penetrate the Veil of the Future, and to discover whether any part of man survives the change we call death; and, if so, how proof of this "survival" is to be obtained. In thus summarizing the historic speculations upon this subject, and in presenting a clear idea of the latest scientific attempts to solve the mystery—by Psychical Research—I have, I trust, afforded the reader a clearer conception of the problem involved than he, perhaps, previously had held.

The problem of Death—the Mystery of the Ages—the Riddle of the Sphinx—is likely to remain unsolved, so long as Science deals with *phenomena* merely; only when we deal with *noumena*—the innermost nature of things-in-themselves—will this problem ever be solved, in the true sense of the word. Life and Death are themselves noumenal things, and exist in a noumenal world. Their ultimate essence or nature can thus only be solved by methods which deal directly with noumena,

—which physical science does not and never can do. All that science can do is to study phenomena. Yet even these may yield us facts of considerable importance —may even yield us a key to the nature of the noumena themselves. It is my hope that the present book, by drawing attention to some of these phenomena (connected with death) and by indicating how they may be studied,—by other, more competent investigators,—may serve to throw some light upon the nature of this great mystery—DEATH.

APPENDICES

APPENDIX A

On "Vampires"

For several centuries there has existed a belief in *vampires* in certain parts of the world, especially in Silesia, in Moravia, and along the frontier of Hungary. Even to this day such stories are circulated among the people and implicitly credited by them. It is asserted that certain persons, who have died, have the power of returning from time to time (generally at night) and sucking the blood of living persons; and that in this manner they are enabled to maintain themselves in a state, if not of life, certainly one very different from death. They are supposed to be enabled to maintain this sort of intra-cosmic existence so long as they can find fresh blood with which to supply themselves. Preferably they attack young persons who are full-blooded and possess an abundance of vitality. Occasionally these persons wake during the process, and frightful have been some of the fights that are said to have taken place between mortal and vampire! Sometimes one and sometimes the other would be victor. More commonly, however, the person so attacked would not wake, and then he or she would rise in the morning pale, emaciated, weak, and exhausted, for no apparent reason. This went on, as a rule, until that person died, when another would be attacked in like manner. This would continue until the vampire was finally caught, exhumed, his head cut off, and his heart cut out or impaled; when with a fearful shriek, he would finally give up the ghost. When the body of the vampire was impaled, fresh blood would gush out. The body would be so full of blood, on occasion, that it could scarcely contain it all; and it would be found dripping from the lips, or even exuding from the eyes, ears, and pores of the skin! Any one bitten by a vampire would become one himself when his turn came to die. Such was the gruesome belief held for several hundred years in parts of Europe, and which is even yet not extinct.

In a curious old work entitled *The Phantom World,* its author, Augustine Calmet, a priest, attempted to find a rational explanation of these stories, and his ingenious speculations will be found in vol. ii. of that treatise. He says:—

"I lay down at first this principle, that it may be that there are corpses which, although interred some days, shed fluid blood through the pores of their body" (p. 41).

Although this can hardly be said to to be so, it is almost the case, on occasion, as may be seen by referring to the discussion under "Putrefaction." However, our author goes on:—

"I add, moreover, that it is very easy for certain people to fancy themselves sucked by vampires, and that the fear caused by that fancy should make a revolution in their frame sufficiently violent to deprive them of life."

The author was evidently well aware of the power of "suggestion"! Returning to the orginal theme, however, he continues:—

"I now come to those corpses full of fluid blood, and whose hair, beard, and nails had grown again. One may dispute these parts of the prodigies, and be very complaisant if we admit the truth of a few of them. All philosophers know well enough how much the people, and even certain historians, enlarge upon things which appear but a little extraordinary. Nevertheless, it is not impossible to explain their cause physically. Experience teaches us that there are certain kinds of earth which will preserve dead bodies perfectly fresh. The reasons for this have been often explained without my giving myself the trouble to make a particular recital of them. . . .

As to the growth of the nails, the hair, and the beard, it is often perceived in many corpses. While there yet remains a good deal of moisture in the body, it is not surprising that some time we see some augmentation in those parts which do not demand a vital spirit. . . .

"The fluid blood flowing through the canals of the body seems to form a greater difficulty, but physical reasons may be given for this. It might very well happen that the heat of the sun, warming the nitrous and sulphurous particles which are formed in those earths that are proper for preserving the body, those particles having incorporated themselves in the newly-interred corpses, ferment, decoagulate, and melt the curdled blood, render it liquid, and give it the power of flowing by degrees through all the canals. . . . As to the cry uttered by the vampires when the stake is driven through their heart, nothing is more natural; the air which is there confined, and thus expelled by violence, necessarily produces that noise in passing through the throat. . . ."

The figures of the vampires that were seen, Calmet considers to be apparitions, occasionally helped out by dreams and other morbid phenomena. Considering the fact that the author lived and wrote in 1751, his speculations may be considered quite remarkable.

Vampires are, however, not unknown in these days. In an article on "Vampires" in *The Occult Review*, June 1908, Dr. Hartmann described a method of what might be termed vampirage. He said:—

"In the Bible it is claimed that when David grew old, a young girl was given to him to supply him with vitality; and not very many years ago certain institutions based upon this principle were existing in France. Young girls were supplied to old men or women as bedfellows. Usually the old person (after having had to submit to certain precautionary measures) had to sleep between two girls, a fair-haired and a dark one, for which privilege he had to pay a certain sum. All of these girls soon lost vitality, some of them died, and these establishments were finally closed by order of the police."

If popular opinion is correct in its assumption that vitality may thus be transferred from the young to the old, it is easy to understand that there may be a great deal of truth in the theory that it is not healthy for children to sleep in the same bed with old persons. Certainly medical science has given at least tacit approbation to this opinion.

Dr. Hartmann also cites the case of the "wonder girl" at Radein, who attracted considerable attention some time ago. For seven years, according to the statement of the investigators, this girl lived without food or drink, and yet was able to maintain phenomenally good health! According to Dr. Hartmann's theory, she lived upon the vitality of others.

> "Instead of taking food," he says, "she withdrew vitality from the children who were brought to her for the purpose of receiving her blessing. Some of these children sickened, some wasted away and died. She did not do this consciously and willingly, for she was a very pious person, and, owing to her long fasting, even considered a saint."

There are historical records of many similar cases, but nearly all of them have been proved to be fraudulent; and at the present day orthodox science refuses to accept many of them as authentic.

APPENDIX B

The Death of Cells

FROM *Age, Growth, and Death,* by C. S. Minot, pp. 75–76:—

I. DEATH OF CELLS

1. *Causes of Death.*

A. External to the organism:—

1. Physical (mechanical, chemical, thermal, etc.).
2. Parasites.

B. Changes in intercellular substances (probably primarily due to cells):—

1. Hypertrophy.
2. Induration.
3. Calcification.
4. Amyloid degeneration (infiltration).

C. Changes inherent in cells.

2. *Morphological Changes of Dying Cells.*

A. Direct death of cells:—

1. Atrophy.
2. Disintegration and reabsorption.

B. Indirect death of cells:—

1. Necrobiosis (structural change precedes final death).
2. Hypertrophic degeneration (growth and structural change often with nuclear proliferation precede final death).

3. *Removal of Cells.*

A. By mechanical means (sloughing or shedding).
B. By chemical means (solution).
C. By phagocytes.

II. Indirect Death of Cells.

A. Necrobiosis:—

1. Cytoplasmic changes:—
   (*a*) Granulation.
   (*b*) Hyaline transformation.
   (*c*) Inhibition.
   (*d*) Desiccation.
   (*e*) Clasmatosis.
2. Nuclear changes—
   (*a*) Karyorhexis.
   (*b*) Karyolysis.

B. Hypertrophic degeneration:—

1. Cytoplasmic—
   (*a*) Granular.
   (*b*) Cornifying.
   (*c*) Hyaline.
2. Paraplasmic—
   (*a*) Fatty.
   (*b*) Pigmentary.
   (*c*) Mucoid.
   (*d*) Colloid, etc.
3. Nuclear (increase of chromatin).

APPENDIX C

Superstitions, Sayings, etc., concerning Death

"Superstitions, Sayings, etc., concerning Death," by C. W. J., in Chambers' *Book of Days*, vol. ii., pp. 52–53:—

"If a grave is open on Sunday, there will be another dug in a week.

"This I believe to be a very narrowly limited superstition, as Sunday is generally a favourite day of funerals among the poor.

"If a corpse does not stiffen after death, or if the *rigor mortis* disappears before burial, it is a sign that there will be a death in the family before the end of the year.

"In the case of a child of my own, every joint of the corpse was as flexible as in life. I was perplexed at this, thinking that perhaps the little fellow might, after all, be in a trance. While I was considering the matter, I perceived a bystander looking very grave and evidently having something on her mind. On asking her what she wished to say, I received for an answer that, though she did not put any faith in it herself, yet people did say that such a thing was a sign of another death in the family within the twelvemonth.

"If every remnant of Christmas decoration is not cleared out of church before Candlemas Day (February 2), there will be a death that year in the family occupying the pew where a leaf or berry is left.

"An old lady (now dead) whom I knew was so persuaded of the truth of this superstition that she would not be content to leave the clearing of her pew to the constituted authorities, but used to send her own servant on Candlemas Eve to see that her own seat, at any rate, was thoroughly freed from danger.

"Fires and candles also afford presages of death—coffins flying out of the former, and winding-sheets guttering down from the latter. A winding-sheet is produced from a candle; if, after it has guttered, the strip which has run down, instead of being absorbed into the general tallow, remains unmelted; if, under these circumstances, it curls over away from the flame, it is a presage of death to the person in whose direction it points.

"Coffins out of the fire are hollow, oblong cinders, spirited from it, and are a sign of coming death in the family. I have seen cinders which have flown out of the fire picked up and examined to see what they presaged; for coffins are not the only things that are thus produced. If the cinder, instead of being oblong, is oval, it is a cradle, and predicts the advent of a baby; while, if it is round, it is a purse, and means prosperity.

"The howling of a dog at night under the window of a sick-room is looked upon as a warning of death being near.

"Perhaps there may be some truth in this notion. Everybody knows the peculiar odour which frequently precedes death, and it is possible that the acute nose of the dog may perceive this, and that it may render him uneasy; but the same can hardly be alleged in favour of the notion that the screech of an owl flying past signifies the same, for if the owl did scent death and was in hopes of prey, it is not likely that it would screech and so give notice of its presence."

BIBLIOGRAPHY[1]

Alden, Henry M. *A Study of Death.* New York, 1895.

Allen, F. D. *Remarks on the Dangers and Duties of Sepulture; or, Security for the Living,* etc. Boston, U. S. A., 1873.

Anon. *An Essay on Vital Suspension: being an attempt to investigate and ascertain those diseases in which the principles of life are apparently extinguished.* London, 1791.

—— *An Unrealized Duty.* London, 1905.

—— *Book of the Craft of Dying.* (O. E., Anon.) New York, 1917.

—— *Burials Amendment Act.* London. N. D.

—— *Churchyard Literature.* New York, 1881.

—— *Circular of the London Association for the Prevention of Premature Burial.* London. N. D.

—— *Directions for Recovering Persons Apparently Dead from Drowning, and from Disorders occasioned by Cold Liquors.* Philadelphia. N. D.

—— *Dying Pillows of Infidels and Christians.* London, 1889.

—— *Earth-to-Earth Burial, and Cremation by Fire.* London, 1890.

—— *How Burying Alive May be Prevented.* London. N. D.

—— *In Dread of Premature Burial.* London. N. D.

—— *Prayers for the Dead. By an Anglican Priest.* London, 1885.

—— *Premature Burial and its Prevention.* London. N. D.

—— *Premature Burial; with Sir B. W. Richardson's Signs and Proofs of Death.* London, 1908.

[1] The Bibliography refers almost exclusively to the physiological side of the question, as treated in Part I. Books on psychical research, or general works on science, are not given, as being too general for a work of this character.

Anon. *Premature Death.* New York, 1878.

—— *The Cause of Death from Old Age now Revealed to Man.* London, 1883.

—— *The Uncertainty of the Signs of Death.* Dublin, 1748.

Arnold, Sir Edwin. *Death—and Afterwards.* New York, 1897.

Besant, Annie. *Death and After?* New York, 1906.

Bichat. *Recherches physiologiques sur la Vie et la Mort.* Paris, 1880.

Björklund, Gustav. *Death and Resurrection, from the Point of view of the Cell Theory.* Chicago, 1910.

Bostwick, Homer, M.D. *An Inquiry into the Cause of Natural Death; or, Death from Old Age.* New York, 1851.

Bouchut, E. *Traité des Signes de la Mort et des Moyens de ne pas être Enterré Vivant.* Paris, 1849.

—— *Mémoirs sur plusieurs nouveaux Signes de la Mort,* etc. Paris, 1867.

Bordeau, L. *Le Problème de la Mort.* Paris, 1900.

Braid, James. *Observations on Trance; or, Human Hibernation.* 1850.

Brouardel, P. *Death and Sudden Death.* New York, 1896.

Browne, O. *On the Care of the Dying.* London, 1894.

Bruhier, d'A. *The Uncertainty of the Signs of Death,* etc. 1751.

Budge, E. Wallis. *The Book of the Dead.* 3 vols. London, 1913.

Burdett, H. C. *The Necessity and Importance of Mortuaries for Towns and Villages, with Suggestions for their Establishment.* London, 1880.

Burke, William. *On Suspended Animation,* etc. London, 1805.

Calhoun, T. *An Essay on Suspended Animation.* Philadelphia, 1823.

Carpenter. *The Drama of Love and Death.* New York, 1912.

Carrington, Hereward. *Vitality, Fasting and Nutrition.* New York, 1908.

Carrington, Hereward. *Death Deferred.* Philadelphia, 1912.

Chantourelle. *Paper at the Royal Academy of Medicine of Paris, on the Danger of Premature Burial,* etc. 1827.

Child. *Senescence and Rejuvenescence.* Chicago, 1915.

Cobb, Augustus G. *Earth-Burial and Cremation.* New. York, 1892.

Coleman, Edward. *Dissertation on Suspended and Natural Respiration.* London, 1802.

Conaro. *The Art of Living Long.* 1903.

Cooper, M. *Uncertainty of the Signs of Death, precipitate interment and dissection, and funeral solemnities.* London, 1746.

Curry, James. *Popular Observations on Apparent Death from Drowning, Suffocation,* etc. London, 1793.

Davis, A. J. *Death and After Life.* Boston, 1865.

Davis, Mary F. *Death in the Light of the Harmonial Philosophy.* Boston, 1876.

Dastre, A. *Life and Death.* New York, 1911.

De Laurence, Dr. L. W. *The Book of Death and Hindoo Spiritism.* Chicago, 1908.

Deschamps, M. H. *Précis de la Mort apparante.* Paris, 1841.

Desmaire, Paul. *Les Morts Vivants.* Paris, 1862.

Drelincourt, C. *The Fears of Death,* etc. London, 1797.

Duffield. *The Burial of the Dead.* New York, 1887.

Eaves, A. *The Mastery of Death.* London, 1904.

Evans, De Lacy. *How to Prolong Life; an Inquiry into the Cause of Old Age and Natural Death.* London, 1880.

Figuier, Louis. *The To-morrow of Death.* Boston, 1873.

Finot, Jean. *Philosophy of Long Life.* New York, 1910.

Fletcher, M. R. *One Thousand Buried Alive by their Best Friends.* Boston, 1890.

Foote, G. W. *Infidel Death Beds.* London, 1886.

Fothergill, Anthony. *Inquiry into the Suspension of Vital Action in Drowning and Suffocation.* Bath, 1794.

Francis, J. R. *Encyclopedia of Death,* 3 vols. Chicago, 1894.

Franks, John. *Observations on Animal Life and Apparent Death.* London, 1790.

Fuller, John, M.D. *Some Hints relative to the Recovery of Persons Drowned and apparently Dead.* London, 1784.

Gardner, John., M.D. *Longevity.* Boston, 1875.

Gaze, Harry. *How to Live Forever.* Chicago, 1904.

Gerhard, Calvin S. *Death and the Resurrection: an Inquiry into their true nature.* Philadelphia, 1895.

Giles, Alfred E. *Funerals, Suspended Animation, Premature Burials.* Boston, 1895.

Greene, C. *Death and Sleep: their Analogy,* etc. London, 1904.

Hadwen, Dr. Walter B. *The Need of Legislation concerning Burial Reform.* London, N.D.

Haeckel, Ernst. *The Wonders of Life.* New York, 1905.

Hamilton, Robert. *Rules for Recovering Persons Apparently Drowned.* London, 1785.

Harris, J. *Post Mortem Handbook.* New York, 1887.

Harrison, J. B. *The Medical Aspects of Death.* London, 1852.

Hartmann, Franz. *Buried Alive.* Boston, 1895.

Hawes, Dr. *On the Duty of the Relations of those who are in Dangerous Illness, and the Hazard of Hasty Interment.* 1703.

Hicks, A. B. *Hints to Medical Practioners concerning granting Certificates of Death.* London, 1895.

Hooker, J. Stenson, M.D. *Premature Burial and its Prevention.* London. N.D.

Howard, N. *Life in Death,* etc. New York, 1872.

Hufeland, C. W. *Ueber die Ungewissheit des Todes,* etc. 1791.

Humphrey, G. M. *Old Age.* Cambridge, 1889.

Icard, S. *Le Mort réele et la Mort Apparante.* Paris, 1897.

Jacobs, Joseph. *The Dying of Death.* New York. N.D.

James, J. Brindley, M.R.C.S. *Trance: its Various Aspects and Possible Results.* London. N.D.

—— *Death and its Verification.* London, 1908.

Jennings, H. S. *Life and Death in Unicellular Organisms.* Boston, 1920.

Johnson, Alexander. *A Collection of Cases Proving the Practicability of Reviving Persons Visibly Dead,* etc. London, 1773.

—— *Relief from Accidental Death; or, Summary Instructions for the General Institution proposed in* 1773. London, 1785.

Johnston, Hugh. *Beyond Death.* New York, 1903.

Kay, J. P. *The Physiology, Pathology and Treatment of Asphyxia, including suspended animation in new-born children, and from hanging, drowning, wounds in the chest, mechanical obstruction of the air passages, respiration of gases, deaths from cold,* etc. London, 1834.

Kirkland, W. *The New Death.* Boston, 1918.

Laborde, J. V. *Le Traitement physiologique de la Mort.* Paris, 1894.

L. E. A. *Sunset Glories: Quotations from Dying Words.* London, 1886.

Le Bon, G. *Recherches expérimentales sur les Signes diagnostiques de la Mort et sur les Moyens de Prévenir les Inhumations Précipitées.* Paris. N.D.

Lenormand, L. *Des Inhumations Précipitées.* Macon, 1843.

Lorand, Arnold. *Old Age Deferred.* Philadelphia, 1910.

Mackenna, Robert W., M.D. *The Adventure of Death,* New York 1917.

Malcolm, John D. *The Physiology of Death from Traumatic Fever.* London, 1893.

Marvin, F. R. *The Last Words of Distinguished Men and Women.* Troy, U. S. A., 1800.

Maze. *Signes de la Mort.* Paris, 1892.

Mercer, Edward. *Why Do We Die?* New York, 1919.

Maeterlinck. *Death.* New York, 1912.

Metchnikoff, Elie. *Old Age (Smithsonian Reports).*

—— *The New Hygiene.* Chicago, 1907.

—— *The Nature of Man.* New York, 1905.

—— *The Prolongation of Life.* New York, 1908.

Minot, Chas. S. *Age, Growth, and Death.* New York, 1908.

Mitchell, C. P. *Dissolution, Evolution and Medicine.* London, 1888.

Munk, W. *Euthanasia: Medical Treatment in aid of Easy Death.* London, 1887.

Orfila, F. *Directions for the Treatment of Persons who have taken Poison, and those in a State of Apparent Death.* London, 1818.

Paris, Thomas, M.D. *The Post-Mortem Handbook,* New York, 1887.

Peebles Dr. J. M. *Death Defeated,* etc. Battle Creek, Mich., 1908.

Philip, A. P. W. *An Inquiry into the Nature of Sleep and Death.* London, 1834.

Ransome, A. *On the Distribution of Death and Disease.* London, 1884.

Reid, J. *The Philosophy of Death.* London, 1841.

Sabatier, A. *Essai sur la Vie et la Mort.* Paris, 1892.

Savory, Wm. S. *On Life and Death.* London, 1863.

Schopenhauer, A. *Ueber den Tod,* etc. Leipzig, 1886.

Shaler, N. S. *The Individual: a Study of Life and Death.* New York, 1902.

Smyth, N. *The Place of Death in Evolution.* London, 1897.

Snart, John. *An Historical Inquiry Concerning Apparent Death and Premature Interment.* London, 1824.

—— *Thesaurus of Horror; or, the Charnel House Explored.* London, 1817.

Stephens, Dr. C. A. *Natural Salvation.* Privately printed, 1907.

Tebb and Vollum. *Premature Burial.* London, 1896.
Tebb, Wm. *Premature Burial: a Reply to Dr. David Walsh.* London. N.D.
Teichmann, Dr. E. *Life and Death.* Chicago, 1906.
Thoms, William J. *Human Longevity: its Facts and Fictions.* London, 1873.
Thwing, E. P. *Euthanasia in articulo mortis.* New York, 1888.
T. M. E. *Final Triumph: Dying Sayings.* London, 1893.
Tozer, Basil. *Premature Burial and the only True Signs of Death.* London. N.D.
Voronoff, S. *Life: A Study of the Means of Restoring Vital Energy and Prolonging Life.* New York, 1920.
Walker, G. A. *Gatherings from Graveyards,* etc. London, 1830.
Walsh, David. *Premature Burial.* New York, 1898.
Weber, Fred P. *Aspects of Death, and Correlated Aspects of Life.* New York, 1918.
Weismann, A. *Ueber Leben und Tod.* Jena, 1884.
Welby, Horace. *Mysteries of Life, Death, and Futurity.* 1861.
Welchman, E. *Observations on Apparent Death from Suffocation or Drowning, Choke-Damp, Stroke of Lightning,* etc. New York, 1842.
Whiter, W. *A Dissertation on the Disorder called Suspended Animation.* Norwich, 1819.
Wilder, Alex, M.D. *Burying Alive a Frequent Peril.* London. N.D.
—— *The Perils of Premature Burial.* London, 1895.
Wilkinson, J. J. G. *Greater Origins and Issues of Life and Death.* London, 1885.
Williamson, James R. *How the State can Prevent Premature Burial.* London. N.D.
Wilson, John K. *Death, its Meaning and Results.* 1900.
Zschokke. *Meditations on Death and Eternity.* Boston, 1864.

PERIODICALS

The Burial Reformer. (Devoted to the prevention of premature burial.)

MAGAZINE ARTICLES OF IMPORTANCE

Aldis, C. J. B. On the Danger of Tying up the Lower Jaw immediately after Death. *Lancet,* 1850, ii. 601.

Askew, F. D. Death. *Ks. City Med. Index,* 1896, xvii, 154–58.

Barker, F. C. On Real and Apparent Death. *Clin. Chi. M.,* 1897, xviii, 232–38.

Beaumont, J. W. Natural Death and Natural Sleep. *Med. Clin.,* London, 1862, xxi. 328.

Bourke, M. W. Resuscitation of a Child after Ten Minutes' Total Submersion in Water, etc. 1859, xliii. 103.

Bradnack, F. Death, Its Signs, etc. *Buffalo M. & S. Jour.,* 1889–90, xxix. 667, 718.

Brandon, R. Construction of Houses for the Reception of the Dead, etc. 1847, xvi. 574.

Burns, J. F. The Hour at which Death most usually Occurs. *N. Y. Med. Journ.,* 1890, li. 17.

Clark, T. E. Buried Alive. *Quar. Jour. Psych. Med.,* 1871, v. 87–93.

Coldstream, John. A Case of Catalepsy. *Edin. Med. and Surg. Jour.,* lxxxi. 477.

Connell, J. B. Causes and Modes of Death. *Ks. Med. Jour.,* 1895, vii. 59–64.

Coskery, O. J. Remarks on 288 Deaths, etc. *N. Y. Med. Jour.,* 1884, xi. 395–400.

Dana, C. L. The Physiology of the Phenomena of Trance. *Med. Rec.,* 1881, xx. 85–89.

Davis, M. Hasty Burials. *Sanit. Rec.,* 1876, iv. 261.

Denman, J. Resuscitation after Two Hours' Apparent Death. *Med. Press and Cir.,* 1867, iii. 95.

Douglas, H. G. Recovery after Fourteen Minutes' Submersion. *Med. Gaz.*, 1842, i. 448.

Ducachet, H. W. On the Signs of Death, and the Manner for Distinguishing Real from Apparent Death. *Am. Med. Rec.*, 1822, v. 39–53.

Duckworth, Sir D. On the Cessation of Respiration before Circulation. *Edin. Med. Jour.*, 1898, iii. 145–52.

Dunglison, R. Lecture on Death. *Coll. Clin. Rec.* Philadelphia, 1880, i. 28.

Evans, T. R. Signs of Impending Death. *Tr. M. Sec. Va.* Wheeling, U. S. A., 1885, cviii. 218–22.

Fraser, W. Distinctions between Real and Apparent Death. *Pop. Sci. Monthly*, 1880–81, xviii. 401–8.

Gairdner, W. T. Case of Lethargic Stupor or Trance, extending continuously over more than Twenty-three Weeks, etc. *Lancet*, 1883, ii. 1078; and 1884, v. 56.

Goadby. Death Trance. *Med. Ind.*, 1856, i. 90–99.

Godfrey, E. L. B. Report on the Resuscitation of a Young Girl Apparently Dead from Drowning. *Phil. Med. Times*, 1879, ix. 375.

Huffy, T. S. Two Cases of Apparent Death. *Tr. Med. Soc. N. Car.*, 1874, xxi. 126–31.

Jamieson, W. A. On a Case of Trance. *Edin. Med. Journ.*, 1871–72, xvii. 29–31.

Jenks, W. G. Causes of Sudden Death. *Mass. Med. Jour.*, 1898, xviii. 10–15.

Knapp, B. C. Sudden Death from Affections of the Nervous System. *Tr. Mass. Med. Leg. Soc.*, 1888, ii. 118–22.

Lange, J. C. Modes of Dying. *Pitts. Med. Rev.*, 1889, iii. 27–34.

Lee, W. The Extreme Rarity of Premature Burial. *Pop. Sci. Month.*, xvii. 526.

Mackenzie, S. C. Phenomena Occurring after Death. In *Med. Leg. Exp. in Calcutta.* Edinburgh, 1891.

Mackey, G. E. Premature Burials. *Pop. Sci. Month.*, xvi. 389.

Madden, T. M. On Lethargy or Trance. *Dub. Jour. Med. Sci.,* 1881, lxxi. 297.

—— On Morbid Somnolence and Death Trance. *Med. Mag.,* London, 1897, vi. 857–922.

Mann, T. D. On Sudden Death from Natural Causes. *Lancet,* 1897, i. 1730–33.

Miller, T. C. The State of the Eyelids after Death—Open or Shut? *Med. Rec.,* 1877, xii. 4.

Minot, C. S. Death and Individuality. *Science,* 1884, iv. 398–400.

Myers, F. W. H. How we Feel When we Die. *Jour. Amer. Med. Assoc.* Chicago, 1897, xxix. 275.

Osborne, W. G. Impositions of an Indian Fakeer, who professed to be Buried Alive, etc. *Lancet,* 1839, i. 885.

Philip, A. P. W. On the Nature of Death. *Phil. Tr.,* 1834, cxxiv. 167–98.

Povall, R. An Account of a Successful Resuscitation of three Persons from Suspended Animation by Submersion for 25 minutes. *West. Med. and Phys. Jour.,* 1828, ii. 499–503.

Reid, T. J. A Case of Suspended Animation. *St. Louis Clin. Rec.,* 1879, vi. 261–63.

Report of Committee on Suspended Animation. *Proc. Roy. Med. and Chir. Soc.,* 1862, iv. 142–47; vi. 299, etc.

Richardson, B. W. Researches on Treatment of Suspended Animation. *Brit. and For. Chir. Rev.,* 1863, xxxi, 478–505.

—— The Absolute Signs and Proofs of Death. *Brit. Med. Jour.,* London, 1888, ii. 1338.

Richardson, F. T. Observations upon a Particular Point in the Physiology of Death. *Month. Jour. Med. Soc.* London, 1842, ii. 260–64.

Richardson, J. The Death Odour. *Peoria Mag.,* 1884, v. 300.

Romero, Fr. Infallible Sign of Extinction of Vitality in Sudden Death. *Med. Tr. Roy. Coll. Phys.,* 1815, v. 478–85 (Latin).

Sawhill, W. F. Sudden Deaths. *Ks. Med. Jour.*, 1897, ix. 505–10.

Shrock, N. M. On the Signs that Distinguish Real from Apparent Death. *Transyl. Jour. Med.*, 1835, xiii. 210–20.

Silvester, H. R. A New Method of Resuscitating Still-Born Children, and of Restoring Persons Apparently Drowned and Dead. *Brit. Med. Jour.*, 1858, 576–79.

Taylor, J. Case of Recovering from Hanging. *Glas. Med. Jour.*, 1886, xiv. 387.

Twedell, H. M. Account of a Man submitted to be Buried Alive for a Month at Jaisulmer. *India Med. Jour.*, 1836, i. 389–91.

Van Giessen, R. E. Sudden Death without Assignable Cause. *Bull. N. Y. Path. Soc.*, 1881, 2 series, i. 279–85.

White, W. A. H. A Case of Trance. *Brit. Med. Jour.*, 1884, ii. 52.

Anon. Causes of Apparent Death. *Cal. Med. Jour.*, 1869, ii. 380–87.

—— Death and Work. *Harper's Weekly*, Sept. 26, 1908.

—— Death by Lightning. *Sci. Amer. Suppt.*, Sept. 10, 1910.

—— Fear of Death. *Current Lit.*, Sept. 1909.

—— Instinct of Feigning Death. *Sci. Amer. Suppt.*, June 6, 1908.

—— Is Death Universal? *Jour. Sc.*, London, 1885, vii. 68–72.

—— Medical Conceptions of Death. *Current Lit.*, Oct. 20, 1909.

—— Old Age and Death. *Sci. Amer. Suppt.*, Feb. 20, 1909.

—— Sensations during Anaesthesia and Death. *Current Lit.*, Nov., 1908.

—— Signs of Death. *Lond. Med. Rec.*, 1874, ii, 205, 221.

—— Sting of Death, The. *The Living Age*, July 2, 1910.

THE END

THE LITERATURE OF DEATH AND DYING

Abrahamsson, Hans. **The Origin of Death:** Studies in African Mythology. 1951

Alden, Timothy. **A Collection of American Epitaphs and Inscriptions with Occasional Notes.** Five vols. in two. 1814

Austin, Mary. **Experiences Facing Death.** 1931

Bacon, Francis. **The Historie of Life and Death with Observations Naturall and Experimentall for the Prolongation of Life.** 1638

Barth, Karl. **The Resurrection of the Dead.** 1933

Bataille, Georges. **Death and Sensuality:** A Study of Eroticism and the Taboo. 1962

Bichat, [Marie François] Xavier. **Physiological Researches on Life and Death.** 1827

Browne, Thomas. **Hydriotaphia.** 1927

Carrington, Hereward. **Death:** Its Causes and Phenomena with Special Reference to Immortality. 1921

Comper, Frances M. M., editor. **The Book of the Craft of Dying and Other Early English Tracts Concerning Death.** 1917

Death and the Visual Arts. 1976

Death as a Speculative Theme in Religious, Scientific, and Social Thought. 1976

Donne, John. **Biathanatos.** 1930

Farber, Maurice L. **Theory of Suicide.** 1968

Fechner, Gustav Theodor. **The Little Book of Life After Death.** 1904

Frazer, James George. **The Fear of the Dead in Primitive Religion.** Three vols. in one. 1933/1934/1936

Fulton, Robert. **A Bibliography on Death, Grief and Bereavement:** 1845-1975. 1976

Gorer, Geoffrey. **Death, Grief, and Mourning.** 1965

Gruman, Gerald J. **A History of Ideas About the Prolongation of Life.** 1966

Henry, Andrew F. and James F. Short, Jr. **Suicide and Homicide.** 1954

Howells, W[illiam] D[ean], et al. **In After Days;** Thoughts on the Future Life. 1910

Irion, Paul E. **The Funeral:** Vestige or Value? 1966

Landsberg, Paul-Louis. **The Experience of Death:** The Moral Problem of Suicide. 1953

Maeterlinck, Maurice. **Before the Great Silence.** 1937

Maeterlinck, Maurice. **Death.** 1912

Metchnikoff, Élie. **The Nature of Man:** Studies in Optimistic Philosophy. 1910

Metchnikoff, Élie. **The Prolongation of Life:** Optimistic Studies. 1908

Munk, William. **Euthanasia.** 1887

Osler, William. **Science and Immortality.** 1904

Return to Life: Two Imaginings of the Lazarus Theme. 1976

Stephens, C[harles] A[sbury]. **Natural Salvation:** The Message of Science. 1905

Sulzberger, Cyrus. **My Brother Death.** 1961

Taylor, Jeremy. **The Rule and Exercises of Holy Dying.** 1819

Walker, G[eorge] A[lfred]. **Gatherings from Graveyards.** 1839

Warthin, Aldred Scott. **The Physician of the Dance of Death.** 1931

Whiter, Walter. **Dissertation on the Disorder of Death.** 1819

Whyte, Florence. **The Dance of Death in Spain and Catalonia.** 1931

Wolfenstein, Martha. **Disaster:** A Psychological Essay. 1957

Worcester, Alfred. **The Care of the Aged, the Dying, and the Dead.** 1950

Zandee, J[an]. **Death as an Enemy According to Ancient Egyptian Conceptions.** 1960